MECHANICAL BEHAVIOR AND PROPERTIES OF COMPOSITE MATERIALS

MECHANICAL BEHAVIOR AND PROPERTIES OF COMPOSITE MATERIALS

VOLUME 1

CARL ZWEBEN
H. THOMAS HAHN
TSU-WEI CHOU

Reviewing Editors
LEIF A. CARLSSON, Ph.D.   JOHN W. GILLESPIE, JR., Ph.D.

TECHNOMIC
PUBLISHING CO., INC.
LANCASTER · BASEL

Delaware Composites Design Encyclopedia—Volume 1
a **TECHNOMIC**®publication

Published in the Western Hemisphere by
Technomic Publishing Company, Inc.
851 New Holland Avenue
Box 3535
Lancaster, Pennsylvania 17604 U.S.A.

Distributed in the Rest of the World by
Technomic Publishing AG

Printed in the United States of America
10 9 8 7 6 5 4 3 2 1

Main entry under title:
   Delaware Composites Design Encyclopedia—Volume 1/Mechanical Behavior and
      Properties of Composite Materials

A Technomic Publishing Company book
Bibliography: p.

Library of Congress Card No. 89-51098
ISBN No. 87762-684-7

C O N T E N T S

v

FOREWORD

The Delaware Composites Design Encyclopedia provides users with basic knowledge about the design and analysis of composite materials and structures. The six-volume indexed set is an ongoing series to which new volumes will be added as new needs arise and new knowledge is gained about this rapidly growing field of composites. Unlike many other guides on this subject, the encyclopedia emphasizes the underlying fundamental science base—documenting the high-quality research in the various disciplines that contribute to the field—in addition to engineering solutions to specific problems. It is in-tended for use by engineers, materials scientists, designers, and other technical personnel involved in the applications of composite materials to industrial products.

The material contained in the encyclopedia was written by international experts in the field and compiled at the University of Delaware's Center for Composite Materials (CCM). Established in 1974, CCM began its University-Industry Consortium, "Applications of Composite Materials to Industrial Products," in 1978 to meet the needs of the aerospace, automotive, electronics, and consumer products industries. Now regarded as an international leader in composites research and education, the Center continues to be supported by the Consortium as well as by the National Science Foundation, the United States Army Research Office-University Research Initiative Program, other federal agencies, and the State of Delaware.

The first version of the *Delaware Composites Design Encyclopedia* was published in 1981 and offered as a special benefit to Consortium members. It has since grown in size and scope and is now being offered to the composites community at large for the first time. The current volumes cover the following subjects:

Volume 1 – Mechanical Behavior and Properties of Composite Materials
Volume 2 – Micromechanical Materials Modeling
Volume 3 – Processing and Fabrication Technology
Volume 4 – Failure Analysis of Composite Materials
Volume 5 – Design Studies
Volume 6 – Test Methods
Index to Volumes 1–6

Dr. John W. Gillespie, Jr. and Dr. Leif A. Carlsson, review editors of the *Delaware Composites Design Encyclopedia*, have incorporated the most recent references relevant to the subject matter to maintain the highest quality, up-to-date encyclopedia of this type.

Volume 1, authored by international experts Carl Zweben, H. Thomas Hahn and Tsu-Wei Chou, begins with a basic introduction to roles and properties of the constituents and their relationship to the mechanics of composites. These sections provide the background information for more detailed subject matter on static strength and fatigue and elastic properties of continuous, discontinuous and woven fabric composites. By emphasizing the fundamentals that affect material performance, engineers and scientists will be able to select the types of fibers, matrices and fiber architectures that will best fulfill the processing and performance requirements of the composites application.

John W. Gillespie, Jr., Ph.D.
Review Editor and Assistant
Director for Research
Center for Composite Materials
University of Delaware
Newark, Delaware 19716 USA

SECTION 1.1

Introduction

This volume is intended to help the designer to answer the following questions:

- Which composite material system should I use?
- How do the properties of composites differ from those of metals and unreinforced plastics?
- What are the properties of the most important composites?
- What are the phenomenological aspects of fatigue properties of composites?

Solids are commonly divided into four classes: metals, ceramics, glasses and polymers [1]. Fibrous composites have been made using fibers and matrix materials from all of these groups of materials. The most important category of composites at present uses polymers for the matrix phase combined with graphite, glass, ceramic or polymeric fibers. The scope of this section is limited to these materials. As fillers are frequently used in polymeric-matrix composites, these constituents will also be considered.

1.1.1 General Characteristics of Composite Materials

Resin-matrix composites are a rapidly expanding class of materials made by combining polymeric matrices with various forms of fibrous reinforcements. Particulates and other types of fillers are sometimes added as well. Composite properties depend on many factors, among which are the properties of the constituents; the form of fibrous reinforcement used (chopped, woven, etc.); fiber volume fraction, length, distribution and orientation; interfacial bond strength; and void content. Many of these factors depend strongly on the fabrication process used. Therefore, it is important that values used in design reflect the process by which the structural component is manufactured.

Because of the vast array of composite materials available, it is difficult to generalize when discussing mechanical properties. However, there are some common features: composites are all heterogeneous, most are anisotropic, and they generally display considerable properties' variability compared to most structural metallic alloys.

Heterogeneous composites have properties that vary from point-to-point throughout the material. For example, when a point inside the material is selected at random, the properties at that point can be very different depending on whether it falls in the matrix or one of the fibers. While it is true that all materials are heterogeneous on the microscopic level, the degree of heterogeneity is generally more severe for composites than for metals. This is because the two primary phases, fiber and matrix, have radically different properties and geometries.

Heterogeneity contributes to another important composite material characteristic, property variability. As mentioned earlier, composite strength and stiffness characteristics depend strongly on fiber orientation, fiber spatial distribution, and the variability of fiber properties. Since it is impossible to position each fiber individually, there is an inherent property variability in composites. Composites reinforced with discontinuous fibers, such as bulk molding compounds and sheet molding compounds, are particularly prone to property nonuniformity because it is difficult to control local fiber content and orientation in the face of material flow. As a result, material stiffness and strength properties vary from point-to-point throughout the material.

Resin properties are strongly dependent on processing conditions, such as temperature and pressure, which can vary significantly throughout a component during its fabrication. Where two regions of material flow come together, there is a "knit line" which is not crossed by any fibers. This knit line is a weak region and a frequent cause of part failure.

Other major items contributing to property variation throughout a part include the following: the material may contain voids, which serve as local stress concentrations, and the important fiber-matrix interfacial bond may vary, as it depends on many factors, including the level and type of applied surface finish [2].

Anisotropy is another property common to most composites. An anisotropic material is one whose properties vary with direction. For example, a composite reinforced with straight, parallel continuous fibers is significantly stronger and stiffer in the direction parallel to the fibers than it is in the transverse direction [2].

In the following discussion we consider the properties of the important fibers, matrix materials and fillers. We then describe characteristics of composites with different forms of reinforcement. Finally we consider mechanical property data and the tests by which they are determined.

1.1.2 Fibers

General Characteristics of Reinforcing Fibers

There are several major categories of man-made reinforcing fibers: glass, graphite (carbon), organic, boron, and ceramic. This subsection considers the properties of these fibers, emphasizing the most important ones: glass, graphite, and organic.

The most important naturally occurring reinforcing fiber is asbestos. In recent years there have been several studies linking asbestos fibers to health disorders. As a result, many users are attempting to convert their products to other reinforcements. For comparative purposes, we will include the properties of asbestos fibers.

Although their properties are extremely different, man-made fibers have some common characteristics. As a rule, the fibers are rather fine, with diameters ranging from about 2 to 13×10^{-6} m (1 to 5×10^{-4} in). The main exceptions are boron fibers which are available in diameters of 1 to 2×10^{-4} m (4 to 8×10^{-3} in). Man-made fibers are generally produced in continuous processes, although later they may be chopped, cut or milled into short lengths. It is difficult to measure fiber properties directly because they are extremely fine and generally quite brittle. As a rule, the only mechanical properties that are routinely determined are extensional tensile modulus and tensile strength. The breaking load of a particular filament can be determined relatively easily. However, definition of a cross-sectional area, which may vary along the length of the fiber, is not simple, and this can affect reported values for breaking stress (breaking load divided by cross-sectional area) and modulus [3].

The subject of fiber tensile strength is important and deserves some consideration. The major reinforcing fibers have tensile stress-strain curves that are linear to failure. One exception is Kevlar[1] 29 aramid which has a modulus that increases somewhat with increasing stress. The absence of yielding makes fibers sensitive to imperfections, which has two important effects on tensile strength: there is considerable scatter at a given gage length, and mean strength decreases with increasing gage length [4]. Therefore, a reported value of mean fiber strength requires the associated test length to be specified. Further, the amount of scatter is important in evaluating fiber strength. Ideally, fiber strength properties should be measured at several lengths.

The problem of understanding fiber tensile strength is complicated by the fact that fiber tensile strength is measured using single fibers, untwisted bundles of fibers (ends), impregnated ends, twisted ends (yarns), and composite tensile coupons. As a rule, each one gives a different strength value. The reasons for this are discussed in the subsections dealing with the properties of unidirectional composites (composites reinforced with continuous straight, parallel fibers) and test methods. Fiber strength values reported in this subsection are based on single filament tests and unidirectional composite coupon tests. The latter data are of greatest relevance to designers.

The subject of composite impact resistance is extremely complex, and there is no one number that can be used to characterize completely the damage resistance of a given material. However, the area under a fiber stress-strain curve, which is the energy it absorbs in failing, provides some useful information about its contribution to the impact resistance of a composite. When fibers have different compressive and tensile strengths, or when these stress-strain curves are nonlinear as in the case of Kevlar 49 in compression [5], compressive and tensile energies to failure are different.

---

[1]Registered trademark of E. I. du Pont de Nemours & Co., Inc.

Glass Fibers

Glass fibers are the cheapest and most widely used man-made composite reinforcements. They are also the oldest, dating back to the period of World War II. Glass fibers generally have high strength-to-weight ratios, but their elastic moduli, which are in the range of those of aluminum alloys, are low compared to the newer fibers such as graphite and aramid [2]. The internal structure of glass fibers is amorphous (noncrystalline) and they are generally regarded to be isotropic. As fibers comprise only a part of the composite, and as they are generally oriented in several directions, the modulus-to-density ratios of glass fiber-reinforced plastics are substantially lower than those of metals [2]. This is one of their major limitations as a structural material.

The increase of deformation with time when a material is subjected to a constant load is called creep. The creep resistance of glass fibers at room temperature is substantially better than that of plastics, but not as good as that of structural metals like aluminum and steel. The addition of glass fibers to plastics greatly reduces their tendency to creep.

The failure of a material after a period of time under constant load is referred to as creep rupture or static fatigue [1]. As in the case of creep, the properties of glass at room temperature fall between those of plastics and metals, but much closer to the latter [2].

Glass fibers generally have good chemical resistance and are noncombustible. They do not absorb water, but their tensile strength is substantially reduced by the presence of moisture.

The coefficient of thermal expansion of glass is an order of magnitude lower than that of most plastics and is lower than most aluminum and steel alloys [2]. Reduction of coefficient of thermal expansion is an important reason for the addition of glass fibers to plastics for many applications. A material with a low coefficient of expansion is frequently said to be dimensionally stable, but other factors enter into resistance to dimensional changes including creep and swelling from moisture absorption.

The strength, modulus and creep rupture resistance of glass fibers decrease with increasing temperature. Conversely, creep rate increases. However, the useful temperature range is quite large. Glass does not soften substantially until temperatures over 500°C (1000°F) are reached.

Table 1.1-1. Representative properties of E glass and S glass fibers.

	E Glass	S Glass
Density, δ (g/cm³)	2.60	2.50
Young's Modulus, E (GPa)	72	87
Tensile Strength,* σ_T (GPa)	1.72	2.53
Tensile Elongation,* %	2.4	2.9
Specific Modulus, E/δ (MNm/kg)	27.7	34.8
Specific Strength,* σ/δ (MNm/kg)	0.66	1.01
Longitudinal CTE ($10^{-6}/°C$)	5.0	5.6

*In a typical composite.

Glass fibers are extremely sensitive to abrasion damage, and they are coated with protective sizings during their production. The sizings also serve as a lubricant and promote adhesion between the fibers and the resins. It is important to select fibers with sizings that are compatible with the matrix with which they are to be used.

When fibers are to be woven into fabric, they are coated with special textile sizings which are removed by heating or "burned off" after weaving. Subsequently, they are treated with another finish to reduce abrasion damage and promote resin adhesion.

There are two important types of glass reinforcing fibers, E and S. A third type, C glass, is used where corrosion resistance is particularly important [2]. The most widely used reinforcing fiber, by far, is E glass, where the E designates electrical grade. This lime-alumina-borosilicate glass does not have a fixed composition. Producers vary constituents based on raw material costs and process considerations. Within prescribed ranges, variations in glass formulation are not thought to affect mechanical properties substantially. However, the user should be alert to the fact that producers use different compositions from time to time and should assure themselves that these differences do not affect product performance in their applications.

Table 1.1-1 presents properties of E glass fibers. The tensile strength value is based on what is obtained in a representative unidirectional composite. Figure 1.1-1 shows how the mean tensile strength of single filaments varies with gage length [6]. As discussed earlier, mean strength decreases significantly with increasing length. E glass fibers have relatively high energies to failure, providing their composites with relatively good impact resistance compared to those of other fibers.

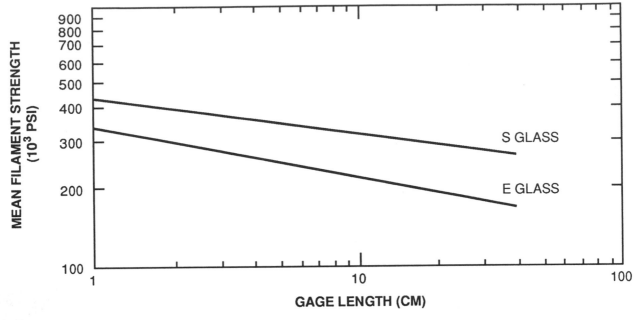

FIGURE 1.1-1.

Composites reinforced with *E* glass fibers are employed in an enormous number of applications. Their uses are so broad that they are generally grouped into several major categories. Some typical examples are listed below:

- automotive: grill opening panels, fender extenders, cab and body components
- agricultural: tractor hoods, fenders, seats, feed troughs
- appliances: air conditioner cases, fans, washing machine tubs, gears
- aerospace: radomes, aircraft fairings and interiors
- business machines: housing, gears, circuit boards, frames
- chemical: pipes, tanks, ducts, filters
- construction: concrete forms, translucent skylights, curtain walls
- electrical/electronic: electric pole crossarms, insulators, electronic components
- marine: boats, water tanks, barge covers
- recreational: motor homes, trailers, campers, snowmobiles
- transportation: truck trailer panels, rapid transit car ends, seating

S glass is a high-strength glass initially developed for military applications. Its modulus is about twenty per-cent greater than that of *E* glass and it is about one-third stronger. Table 1.1-1 presents the properties of this material. As for *E* glass, the reported value of tensile strength is based on composite strength data. Figure 1.1-1 shows the decrease in mean filament strength with increasing gage length.

The failure energy of *S* glass fibers is high, and the impact resistance of composites made from them is among the highest of all fiber-reinforced materials. The creep rupture resistance of this material is significantly better than that of *E* glass. Despite its generally better properties, the use of *S* glass is far more limited than *E* glass because of its higher cost.

Graphite (Carbon) Fibers

Graphite fibers, sometimes referred to as carbon fibers, comprise one of the most important classes of reinforcement with enormous potential for future growth. Their primary advantages over glass fiber are higher modulus, lower density, much better fatigue properties, improved creep rupture resistance, and lower coefficient of thermal expansion [2]. Creep at room temperature is generally considered to be negligi-ble. On the negative side, because of their low strain-to-failure ratio, failure energies are relatively low. As a result, the impact resistance of graphite fiber composites is generally lower than that of glass fiber composites. However, it should be emphasized that the subject of impact resistance is very complex, and generalizations can sometimes be misleading.

Graphite fibers of commercial importance are currently made from three precursors: polyacrylonitrile (PAN) fiber [7], rayon fiber, and petroleum pitch [8], a residue of the refining process. Production of high-modulus graphite fibers by pyrolysis of rayon and PAN fibers dates back to the 1960s. (Low-modulus carbon fibers produced by pyrolysis of rayon cloth, used for re-entry vehicles and rocket nozzles, were introduced in the 1950s.) Spinning of graphite fibers from pitch is a more recent development. Some believe that, because of lower raw material costs, pitch fibers are likely to be less expensive than those using fiber precursors. However, to date, pitch fibers have been able to match the moduli that can be obtained with PAN- and rayon-based fibers, but their strength properties have been substan-

tially lower. The high strength and moduli of graphite fibers, regardless of precursor, result from their high degree of crystallinity and orientation. Because of their high degree of internal structure orientation, graphite fibers are strongly anisotropic. Transverse extensional modulus and shear moduli for most fibers are generally an order of magnitude lower than axial modulus [9].

The most important graphite fibers at the present time are PAN-based, and these will be considered in this section. There are many graphite fibers on the market. They can be divided, somewhat arbitrarily, into three major categories: high-strength, high-modulus and ultrahigh-modulus. Table 1.1-2 presents representative properties of these materials. The tensile strengths reported are effective values that would be obtained in representative unidirectional composite specimens. Figure 1.1-2 shows the considerable amount of scatter and decrease in mean strength for high-modulus graphite fibers reported by Diefendorf and Tokarsky [10]. According to the authors, this results from local

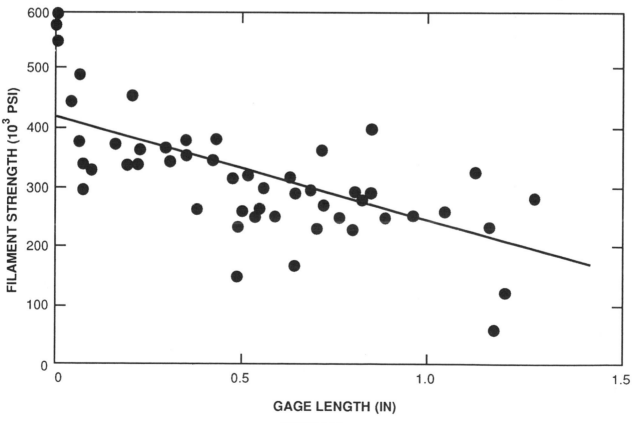

FIGURE 1.1-2.

Table 1.1-2. Representative properties of graphite (carbon) fibers.

	High-Strength	High-Modulus	Ultrahigh-Modulus
Density, δ (g/cm³)	1.8	1.9	2.0–2.1
Young's Modulus, E (GPa)	230	370	520–620
Tensile Strength,* σ_T (GPa)	2.48	1.79	1.03–1.31
Tensile Elongation,* %	1.1	0.5	0.2
Specific Modulus, E/δ (MNm/kg)	128	195	260–295
Specific Strength,* σ_T/δ (MNm/kg)	1.38	0.94	0.52–0.62
Longitudinal CTE ($10^{-6}/°C$)	−0.4	−0.5	−1.1**

*In a typical composite.
**Estimated.

microcompressive buckling imperfections that are created during the cool down phase of fiber production. High-modulus fibers are more prone to this kind of imperfection because of their greater internal orientation.

As the tensile stress-strain curve of graphite fibers is linear to failure, ultimate strain is simply equal to tensile strength divided by modulus. Since tensile strength decreases with increasing modulus, the failure strain of high-modulus fibers is considerably lower than that of high-strength fibers. As a result, composites made from high-modulus and ultrahigh-modulus fibers tend to be more brittle than those reinforced with high-strength fibers. This makes them more sensitive to stress concentrations such as those that occur around joints and cutouts.

Table 1.1-3. Representative properties of Kevlar 29 and Kevlar 49 aramid fibers.

	Kevlar 29	Kevlar 49
Density, δ (g/cm³)	1.44	1.44
Young's Modulus, E (GPa)	83	124
Tensile Strength,* σ_T (GPa)	2.27	2.27
Tensile Elongation,* %	2.8	1.8
Specific Modulus, E/δ (MNm/kg)	57.6	86.1
Specific Strength,* σ/δ (MNm/kg)	1.58	1.58
Longitudinal CTE ($10^{-6}/°C$)	–	−2

*In a typical composite.

Organic Fibers

Some natural organic fibers, such as cotton, jute and sisal, are used as reinforcements. However, because of their low modulus, the mechanical properties they provide are modest, and they are of little interest for structural applications. The same is true of synthetic organic fibers, with the exception of aromatic polyamides (aramids).

There are several commercial aramid fibers: Nomex,[2] Kevlar, Kevlar 29 and Kevlar 49. Applications of Nomex include high-temperature fabrics, filters and structural honeycomb for sandwich-core laminates. Kevlar is used for tire cord. The two fibers of interest for reinforcing plastics are Kevlar 49 and, to a lesser extent, Kevlar 29. Table 1.1-3 presents representative properties of these two fibers. While the tensile stress-strain curve of Kevlar 49 is linear to failure, that of Kevlar 29 is slightly concave upward. The modulus of Kevlar 29 increases from an initial value of 83 GPa (12 × 10⁶ psi) to a value at failure of about 100 GPa (15 × 10⁶ psi). Kevlar 49 has an axial modulus of 124 GPa (18 × 10⁶ psi) and a specific gravity of 1.44, about the same as Kevlar 29. Because of its higher stiffness, Kevlar 49 is used more widely as a reinforcement, and the remainder of this subsection will be devoted to a discussion of this fiber.

Figure 1.1-3 shows the tensile strength and strain to failure of Kevlar 49 aramid fibers as a function of gage length [11]. As for glass and graphite, mean strength decreases as test length increases. The line representing ultimate strain roughly parallels that of strength. The slight difference in slope is probably the result of experimental uncertainty in determination of ultimate strains.

The modulus of Kevlar 49 is between that of E glass, 72 GPa (10.5 × 10⁶ psi), and high-strength graphite, 230 GPa (34 × 10⁶ psi). Because its density is much lower than that of E glass (1.44 compared to 2.6), the specific modulus of Kevlar 49, 86 GPa (12.5 × 10⁶ psi), is three times that of E glass, but it is significantly lower than that of high-strength graphite, 130 GPa (19 × 10⁶ psi).

The tensile stress-strain curve of composites reinforced with aligned Kevlar 49 fibers (unidirectional composites) is linear to failure when they are loaded in the fiber direction. In this respect they are similar to the

[2]Registered trademark of E. I. du Pont de Nemours & Co., Inc.

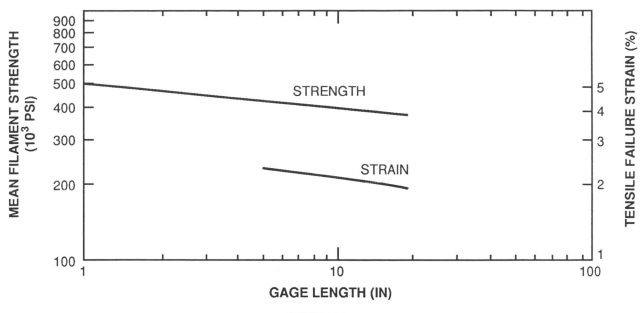

Y-axis (left): MEAN FILAMENT STRENGTH (10^3 PSI)

900
800
700
600
500
400
300
200
100

X-axis: GAGE LENGTH (IN)   1   10   100

Y-axis (right): TENSILE FAILURE STRAIN (%)   5   4   3   2   1

STRENGTH

STRAIN

FIGURE 1.1-3.

other composites discussed in this section. However, their compressive behavior is unique. The compressive stress-strain curve resembles that of an elastic-plastic material [12]. This nonlinearity results from a fiber failure mode which appears to be associated with an instability or kinking of the fiber microstructure. The onset of compressive nonlinearity occurs at a fiber stress of about 440 MPa (65 × 10^3 psi), which corresponds to a strain of 0.35%. Because of these properties, the compressive strengths of composites reinforced with Kevlar 49 aramid fibers are, in many cases, significantly lower than those using other reinforcements such as glass and graphite.

Kevlar 49 fibers have excellent tensile-tensile fatigue resistance. Their compressive fatigue characteristics have not been studied extensively, but appear to be good within the proportional limit. The resistance to creep of Kevlar 49 fibers is significantly better than that of other organic fibers. Room temperature creep rates as a function of stress are of the same order of magnitude as those of glass and should be considered in design. The creep rupture resistance of Kevlar 49 falls between those of S glass and graphite.

Kevlar 49 fibers are strongly anisotropic, a property they share with graphite. Transverse extensional modulus and shear moduli are about an order of magnitude lower than axial extensional modulus. Kevlar 49 has the highest specific strength (strength divided by density)

of any commercial material. As a result it is the material of choice for high-performance rocket motor cases and pressure vessels.

The tensile energy to failure of Kevlar 49 is much greater than that of high-strength graphite, about the same as E glass and significantly lower than S glass. Although the area under the compressive stress-strain curve of Kevlar 49 is very large, the energy to the proportional limit is relatively low compared to other fibers. Because of these and other factors, the impact behavior of composites reinforced with Kevlar 49 is particularly complex. As a rule of thumb the impact resistance of these materials falls between those of E glass and high-strength graphite composites.

Applications of Kevlar 49 composites include aircraft fairings, wing trailing edges, interiors and other semi-structural applications, high-performance boats and sports equipment such as golf club shafts, skis and tennis rackets. Kevlar 49 is frequently used in combination with other fibers, such as graphite and E glass to obtain a balance of properties and cost that cannot be obtained with a single fiber.

1.1.3 Resins

In this subsection, we consider the mechanical and physical properties of the most common resins which

Table 1.1-4. Common thermosetting resins.

Alkyds
Diallyl Phthalates (DAP)
Epoxies
Furans
Melamines
Phenolics
Polybutadienes
Polyesters (thermosetting)
Polyimides (thermosetting)
Polyurethanes
Silicones
Ureas
Vinyl Esters

are used with reinforcing fibers. Resins (also called polymers and plastics) are generally divided into two main classes, thermosets and thermoplastics. We consider the general characteristics of these types of resins and discuss their use in composites. For a more detailed discussion of resin chemistry, see section 3.3 in Volume 3.

Thermosets

Thermosetting resins (thermosets) are materials which are cured, or hardened into a permanent shape by an irreversible chemical reaction known as cross-linking. Thermosets may soften at elevated temperatures, but they cannot be reformed [13]. Although the first thermoset resins required heat for curing, hence the name, many others in this category undergo cross-linking in the presence of a catalyst, without the addition of heat. Thermosetting polyesters are a prime example. In the curing process, linear polymer chains are joined to form a rigidized three-dimensional structure.

Thermosets are generally brittle and are rarely used without some form of filler or reinforcement. Because of their cross-linked structure, thermosetting plastics have relatively good creep resistance and elevated temperature properties, although modulus and strength do decrease with increasing temperature. The chemical resistance of these materials is relatively good. In particular, thermosetting polyesters and vinyl esters are widely used in applications where corrosion resistance is important.

Table 1.1-4 lists the most important thermosetting resins. The major polymers used in composites are epoxies, vinyl esters and polyesters. Epoxy-matrix composites generally have the best mechanical proper-

ties. These resins are relatively expensive and have long cure cycles, although some more rapid-cure systems have been recently introduced. Polyester composites tend to have elastic properties similar to those of epoxy, but lower strength characteristics. The elastic properties of reinforced vinyl esters are like those of reinforced epoxies and polyesters, and their strengths fall somewhere in between.

Table 1.1-5 presents representative properties of some thermosetting resins. Elastic moduli and strengths are typically an order of magnitude lower than those of metals.

Thermoplastics

Thermoplastic resins are solid at room temperature. They soften or melt when heated and re-harden when cooled [13]. These long-chain polymers do not cross-link as do thermosets, and they can be repeatedly reformed by application of heat, although this may eventually result in material degradation. Table 1.1-6 lists some of the more common thermoplastic polymers.

Thermoplastics generally are tough compared to thermosets and are widely used without reinforcement. However, their stiffness and strength properties, although similar to those of thermosets, are low compared to other structural materials, as Table 1.1-5 shows, so that use of reinforcements is desirable. Thermoplastics can be formed into complex shapes easily and economically by processes such as injection molding, extrusion, and thermoforming. The creep resistance of many thermoplastics, particularly at elevated temperature, is significantly lower than that of thermosets, and this has been a serious impediment to their wider use in structural applications. As a class of materials, thermoplastics are also more susceptible to attack by solvents than are thermosets.

Earlier, thermoplastics were reinforced primarily with discontinuous glass fibers and particulate fillers; however, recently there has been much work with continuous fibers in the area of relatively high-temperature resins like polyetheretherketone, polysulfones and thermoplastic polyimides [14,15]. The elastic and strength properties obtained with composites using these matrices are similar to or better than those employing epoxies, and their impact resistance is significantly better. The good moisture resistance and high temperature range of polyimides makes them particularly attractive, although they are more difficult to process than ep-

Table 1.1-5. Typical resin properties at room temperature (unfilled).

Resin	Type	Density g/cm^3	Tensile Modulus GPa (10^6 PSI)	Tensile Strength MPa (10^3 PSI)
Epoxy	Thermoset	1.1–1.4	2.1–5.5 (0.3–0.8)	40–85 (6–12)
Phenolic	Thermoset	1.2–1.4	2.7–4.1 (0.4–0.6)	35–60 (5–9)
Polyester	Thermoset	1.1–1.4	1.3–4.1 (0.2–0.6)	40–85 (6–12)
Acetal	Thermoplastic	1.4	3.5 (0.5)	70 (10)
Nylon	Thermoplastic	1.1	1.3–3.5 (0.2–0.5)	55–90 (8–13)
Polycarbonate	Thermoplastic	1.2	2.1–3.5 (0.3–0.5)	55–70 (8–10)
Polyethylene	Thermoplastic	0.9–1.0	0.7–1.4 (0.1–0.2)	20–35 (3–5)
Polyester	Thermoplastic	1.3–1.4	2.1–2.8 (0.3–0.4)	55–60 (8–9)
PEEK	Thermoplastic	1.3–1.4	3.5–4.4 (0.5–0.6)	100 (15)
PPS	Thermoplastic	1.3–1.4	3.5 (0.5)	78 (15)

oxies. Polysulfone composites are easier to fabricate than polyimides, but their relatively poor solvent resistance is a severe limitation. PEEK is a relatively new semi-crystalline matrix polymer with excellent solvent resistance and mechanical properties [15].

It appears likely that high-performance thermoplastic-matrix composites will become an important class of high-performance structural materials.

1.1.4 Fillers

Fillers are small organic and inorganic particulate materials mixed with plastics to modify their properties, extend resins in short supply, and reduce cost. There are many fillers available with a variety of shapes: spherical, plate-like, fibrous and irregular. The distinction between short fibers and fibrous particulates is blurred. We arbitrarily define the maximum dimension for a filler to be less than 1 mm (0.04 in). The principal reasons for using fillers are summarized in Table 1.1-7.

Discussion

Many fillers are abundant, naturally occurring materials, which are far cheaper than petroleum-based resins. The dramatic rise in the price of oil has made filler use increasingly attractive for economic reasons. Additionally, occasional resin shortages provide strong motivation for use of fillers to extend resins without adversely affecting product performance.

Filled plastics are *particulate composites* whose properties depend on the following:

- resin properties
- filler properties
- internal geometry, including filler shape, volume fraction, distribution, and orientation (for non-spherical particles)

The wide variety of fillers available makes it possible to modify properties in opposite directions. For example, some fillers produce composites with thermal conductivities that are greater than the resin, others lower. The same holds for electrical conductivity.

Table 1.1-6. Common thermoplastic resins.

Acrylonitrile-butadiene-styrenes (ABS)
Acetals
Acrylics
Cellulosics
Fluoropolymers
Nylons (polyamides)
Polyamide-imide
Polyaryl ether
Polyarylsulfone
Polycarbonates
Polyesters (thermoplastic)
Polyetheretherketone (PEEK)
Polyethersulfones
Polyethylenes
Polyimides (thermoplastic)
Polyphenylene oxides (PPO)
Polyphenylene sulfides (PPS)
Polyphenylquinoxalines (PPQ)
Polypropylenes
Polystyrenes
Polysulfones
Polyvinyl chlorides (PVC)
Styrene-acrylonitriles (SAN)

Table 1.1-7. Reasons for use of fillers.

Economic
- — lower cost
- — resin extension to alleviate shortages

Processing
- — lower exotherm (thermosets)
- — reduced crazing and cracking (because of lower cure shrinkage and thermal expansion)
- — reduced warping
- — thixotropy

Esthetic
- — opacity
- — reduced shrinkage
- — surface quality

Physical Properties Modification
- — lower coefficient of thermal expansion
- — increased electrical resistivity
- — increased electrical conductivity
- — better electromagnetic shielding
- — lower or higher coefficient of friction
- — increased thermal conductivity
- — increased thermal insulation
- — increased flame retardance
- — increased corrosion resistance
- — lower moisture permeability and absorption
- — lower cure shrinkage
- — increased abrasion resistance
- — lower coefficient of friction

Mechanical Properties Modification
- — increased modulus
- — increased hardness
- — increased creep resistance

Table 1.1-8. Fillers and their influence on composite materials.

Calcium Carbonate
- — lower cost; increased modulus and creep resistance
- — lower cure shrinkage, coefficient of thermal expansion and exotherm
- — improved surface quality

Kaolin (Clay)
- — lower cost; increased chemical resistance, electrical resistivity, modulus, and creep resistance
- — lower cure shrinkage, coefficient of thermal expansion and exotherm

Glass Particles
- — increased electrical resistivity, chemical resistance, flame retardance, modulus, hardness, abrasion and creep resistance
- — reduced shrinkage, coefficient of thermal expansion, and exotherm

Mica
- — lower cost; increased electrical resistivity, chemical resistance, modulus, hardness and creep resistance
- — lower thermal conductivity, cure shrinkage, coefficient of thermal expansion, exotherm, moisture absorption, and permeability

Talc
- — lower cost; increased modulus, creep resistance; improved surface quality
- — lower cure shrinkage, coefficient of thermal expansion, and exotherm

Silica
- — lower cost; increased modulus, electrical resistivity, creep resistance, hardness, chemical resistance
- — lower cure shrinkage, coefficient of thermal expansion, and exotherm
- — liquid resin thixotropy

Wollastonite
- — lower cost; increased modulus, hardness, and creep resistance
- — lower cure shrinkage, exotherm, coefficient of thermal expansion, and moisture absorption

Alumina (Aluminum Oxide)
- — increased electrical resistivity and thermal conductivity
- — lower cure shrinkage and coefficient of thermal expansion

Aluminum Trihydrate
- — flame retardance

Antimony Oxide
- — flame retardance

Powdered Metals (Aluminum, Bronze, Iron)
- — increased modulus, creep resistance, thermal conductivity and electromagnetic shielding
- — lower cure shrinkage, coefficient of thermal expansion, and exotherm

Some of the physical property modifications produced by fillers have a beneficial effect on processing. By displacing an equal volume of resin, they reduce the amount of exothermic heat generated in a part during cure, permitting higher processing temperature and lower cycle time. (Too much exothermic heat can raise part temperature enough to cause material degradation.) As the coefficients of thermal expansion of most inorganic fillers are less than those of resins, particulate composites made from these particles also have lower thermal expansion coefficients. This helps to alleviate cracking and crazing during cool down, as does the lower cure shrinkage of composites compared to their parent resins. (Cure shrinkage is a volume reduction that occurs in many resins during cure. It is distinct from the volume change arising from temperature reduction.)

Addition of particulates raises resin viscosity. Use of extremely fine particles produces thixotropy[3] in liquid resins, reducing unwanted flow in fabrication methods, such as hand lay-up. However, increased viscosity is generally undesirable in many processes, like injection molding.

The subject of mechanical properties requires particular consideration. Because they are stiffer than resins, organic fillers generally produce some increase in modulus. Their influence on strength is not as obvious.

Fillers produce internal stress concentrations in particulate composites which are analogous to those occurring in transversely loaded unidirectional composites (see section 1.2). As the composite is loaded, resin-particle interfacial stress increases. If the bond strength is not adequate, separation occurs, and voids are formed. Where this happens, composite modulus decreases, as in the case of discontinuous fiber composites. Further, the presence of stress concentrations and voids tends to reduce material strength. The intensity of these effects depends on factors such as resin toughness and particulate size and volume fraction.

In general, organic particles increase stiffness and improve creep resistance. Depending on bond strength, volume fraction and particle size, they can either improve or degrade strength properties. In either case, composite failure strains are lower than those of the parent resins.

Table 1.1-8 lists many of the fillers in use, along with the properties they impart to composites. According to Seymour [16] calcium carbonate is by far the most widely used, 80%, followed by kaolin (clay), 7½%; starches and cellulosics, 4%; wood flours, 3½%; and glass particles, 3%. There is increasing interest in minerals, such as talc, mica, and wollastonite. Reference [17] provides a detailed discussion of these materials.

It is important to recognize that there is considerable variability in the composition and properties of mineral fillers, depending on their place of origin and processing. The influence of filler impurities and variability on composite properties is a subject that has not been extensively explored, and it deserves further consideration.

[3]A fluid under shear is said to be thixotropic if its apparent viscosity increases with time.

1.1.5 References

1. ASHBY, M. F. and D. R. H. Jones. *Engineering Materials 2—An Introduction to Microstructures, Processing and Design*, Pergamon Press (1988).
2. HULL, D. *An Introduction to Composite Materials*, Cambridge University Press, NY (1985).
3. WILSON, D. W. and L. A. Carlsson. "Mechanical Characterization of Composite Materials," in *Physical Methods of Chemistry*, B. W. Rossiter, J. F. Hamilton and R. C. Baetzold, eds., Interscience Publishers, in press.
4. ROSEN, B. W. and Z. Hashin. "Analysis of Material Properties," in *Engineered Materials Handbook, Vol. 1* (Composites, ASM International, T. J. Reinhart, Tech. Chairman), p. 185 (1987).
5. MAAS, D. R. "Mechanical Properties of Kevlar/SP328," CCM-83-19, Center for Composite Materials, University of Delaware (1983).
6. METCALFE, A. G. and G. K. Schmitz. "Effect of Length on the Strength of Glass Fibers," *ASTM Preprint No. 87* (June 1964).
7. JAIN, M. K. and A. S. Abhiraman. "Conversion of Acrylonitrile-Based Precursor Fibers to Carbon Fibers: Part 1—A Review of the Physical and Morphological Aspects," *J. Materials Science*, 22:278 (1987).
8. ENDO, M. "Structure of Pitch-Based Carbon Fibers," *J. Materials Science*, 23:598 (1988).
9. HASHIN, Z. "Analysis of Properties of Fiber Composites with Anisotropic Constituents," *J. Appl. Mech.*, 46:543 (1979).
10. DIEFENDORF, R. J. and E. Tokarsky. "High-Performance Carbon Fibers," *Polymer Engineering and Science*, 15(3):150 (1975).
11. ZWEBEN, C., W. S. Smith and M. W. Wardle. "Test Methods for Fiber Tensile Strength, Composite Flexural Modulus and Properties of Fabric-Reinforced Laminates," *Composite Materials: Testing and Design—Fifth Conference*, ASTM STP 674, American Society for Testing and Materials, Philadelphia (1979).
12. ZWEBEN, C. "The Flexural Strength of Aramid Fiber Composites," *Composite Materials*, 12:422 (1978).
13. BILLMEYER, F. W. *Textbook of Polymer Science*, 3rd. ed., John Wiley & Sons, Inc., NY (1984).
14. JOHNSTON, N. J. and P. M. Hergenrother. "High Performance Thermoplastics: A Review of Neat Resin and Composite Properties," NASA TM 89104 (February 1987).
15. *Thermoplastic Composite Materials*. L. A. Carlsson, ed., Elsevier (to be published, 1989).
16. SEYMOUR, R. B. "Plastics Additives: An Update," *Plastics Compounding*, p. 41 (January/February 1980).
17. *Handbook of Fillers and Reinforcements for Plastics*. H. S. Katz and J. V. Milewski, eds., Van Nostrand Reinhold Co. Inc., NY (1978).

SECTION 1.2

Mechanics of Composite Materials

1.2 MECHANICS OF COMPOSITE MATERIALS

C. ZWEBEN

1.2.1 Introduction

To design efficient structures, the engineer must have a good understanding of the behavior of the materials used. The characteristics of metals are familiar to most designers, and these are factored into the design process, often implicitly. For example, in designing bolted joints it is common to assume that the applied load is uniformly resisted by each of the bolts. This assumption is based on the fact that most structural metals display considerable plastic deformation before failure which tends to distribute the load evenly among the bolts. Composites generally do not yield plastically. As a result, the designer cannot assume that all of the bolts in a joint share the load equally. In point of fact, because of the absence of plastic deformation, composites are very sensitive to stress concentrations, such as those that arise at joints and other discontinuities [1]. This is one illustration of an important difference in mechanical behavior between composites and metals and its effect on design.

This section examines the elastic and strength properties of composites, and how they are related to the properties of the fibers and matrix materials of which they are composed. It is hoped that through an understanding of the key factors affecting material performance, the designer will be able to select the types of fibers, resins and reinforcement forms that will best fulfill the requirements for the structures under consideration. This knowledge should also benefit those involved with materials development and processing.

The discussion of mechanical behavior is broken down by reinforcement form: continuous, aligned fibers; woven materials; and aligned or randomly distributed discontinuous fibers. The elastic and strength properties of interest are tension, compression and shear. The approach in this section is primarily descriptive, and use of mathematics is kept to a mini-

mum. More extensive analytical representations are presented in Volume 2.

Consideration of mechanical behavior begins with composites reinforced with continuous, aligned fibers. Many of the principles developed here also apply to the other reinforcement forms discussed later.

1.2.2 Unidirectional Composites

Materials reinforced with straight, parallel fibers are usually referred to as unidirectional composites. In this subsection we consider only continuous fibers. The case of discontinuous aligned fibers is treated later. Figure 1.2-1 shows an idealized representation of the type of composite under consideration. Note that the individual fibers here are randomly dispersed throughout the material. In practice, this may not be the case. Glass, aramid and graphite fibers are produced in the form of bundles of fibers called ends. Unless care is taken to disperse the fibers, they may tend to be clumped together in the material. Figure 1.2-2 shows an end view of such a composite. Ends are sometimes intentionally twisted to facilitate handling and reduce fiber damage during processing. Ends may be removed from their packages by pulling them off the end, rather than by pulling tangentially and rotating the package. This procedure results in twist. The presence of twist interferes with fiber dispersion and promotes grouping of fibers.

When fibers are closely packed together, resin may not penetrate completely into the interstices. This void space has a deleterious effect on the integrity of the material. The influence of these voids will be considered later in this subsection. Fiber proximity has another effect. Local stress concentrations in the material increase as fiber spacing decreases, and this effect reduces material strength properties.

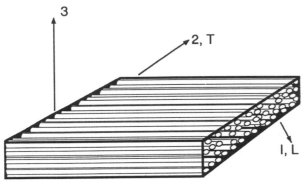

FIGURE 1.2-1.

One of the most important factors affecting composite properties is the amount of fiber it contains. Here, the important parameter is the percentage by *volume* and not by weight. This is an important concept, and one that is frequently misunderstood. To illustrate this point, consider two unidirectional composites made from the same matrix material. Assume that the fibers used in the two materials are identical in every way except that the density of the first is much greater than that of the second. If equal *volumes* of fibers are used in the two composites, their mechanical properties will be identical. However, since the first fibers are more dense than the second, the composite using them has a higher percentage of fibers by *weight*, although the percentages of fiber by *volume* are equal. Conversely, if composites having equal weight percentages are made, the one using the first fibers will have a lower fiber volume fraction and will be weaker and less stiff.

It is very common to find literature references to fiber content that do not state whether the basis is weight or volume. In these cases, it is likely that weight percentage is being reported. A method for converting from weight fraction to volume fraction is described below.

Volume Fraction, Weight Fraction and Density

Fiber volume fraction v_f is defined as the ratio of fiber volume, V_f, to total composite volume, V_c.
That is,

$$v_f = \frac{V_f}{V_c} \tag{1.2-1}$$

Matrix volume fraction v_m is defined as

$$v_m = \frac{V_m}{V_c} \tag{1.2-2}$$

where V_m is the volume of resin in the composite.
Similarly, the volume fraction of voids in the composite is

$$v_v = \frac{V_v}{V_c} \tag{1.2-3}$$

in which V_v is the total volume of voids in the composite.
Since the composite is made up of fibers, matrix, and voids, we have the relation

$$V_c = V_f + V_m + V_v \tag{1.2-4}$$

Dividing by V_c, we obtain an equation relating fiber, matrix, and void volume fraction:

$$v_f + v_m + v_v = 1 \tag{1.2-5}$$

Weight fractions are defined in a similar way. Let the total weights of fiber, matrix and composite be denoted by W_f, W_m and W_c. The corresponding weight fractions are w_f, w_m and w_c. The latter are defined by

$$w_f = \frac{W_f}{W_c} \tag{1.2-6}$$

$$w_m = \frac{W_m}{W_c} \tag{1.2-7}$$

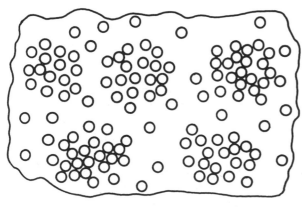

FIGURE 1.2-2.

Since voids have no weight, composite weight is given by

$$W_c = W_f + W_m \qquad (1.2\text{-}8)$$

Dividing by W_c, the following relation is obtained:

$$w_f + w_m = 1 \qquad (1.2\text{-}9)$$

Composite density, ϱ_c, can be related easily to fiber density, ϱ_f, and matrix density, ϱ_m. Using definitions of density:

$$W_f = \varrho_f V_f \qquad (1.2\text{-}10)$$

$$W_m = \varrho_m V_m \qquad (1.2\text{-}11)$$

$$W_c = \varrho_c V_c \qquad (1.2\text{-}12)$$

Using these three relations and Equation (1.2-8), it is found that

$$\varrho_c V_c = \varrho_f V_f + \varrho_m V_m \qquad (1.2\text{-}13)$$

The desired expression is found by dividing this expression by V_c and using Equations (1.2-1) and (1.2-2):

$$\varrho_c = \varrho_f v_f + \varrho_m v_m \qquad (1.2\text{-}14)$$

Composite density is simply the average of fiber and matrix densities weighted by volume fraction. When the matrix consists of resin and filler, the density of the filled resin should be used for ϱ_m.

As discussed previously mechanical properties depend on volume fractions. Since weight fractions are more easily measured, it is useful to be able to relate the two. Dividing the left sides of Equations (1.2-10) and (1.2-11) by W_c and the right sides by the equivalent value, $\varrho_c V_c$, and making use of Equations (1.2-1), (1.2-2), (1.2-6) and (1.2-7), the following are obtained:

$$w_f = \frac{\varrho_f}{\varrho_c} v_f \qquad (1.2\text{-}15)$$

$$w_m = \frac{\varrho_m}{\varrho_c} v_m \qquad (1.2\text{-}16)$$

To eliminate ϱ_c from these expressions, Equation (1.2-14) is used. Then, to obtain the desired form of the

relationships between volume fractions and weight fractions, Equation (1.2-5) is employed. The final expressions are

$$v_f = \frac{w_f \varrho_m (1 - v_v)}{\varrho_f + w_f (\varrho_m - \varrho_f)} \qquad (1.2\text{-}17)$$

$$v_m = \frac{w_m \varrho_f (1 - v_v)}{\varrho_m + w_m (\varrho_f - \varrho_m)} \qquad (1.2\text{-}18)$$

Equations (1.2-17) and (1.2-18) allow a direct determination of volume fractions when weight fractions, component densities and void volume fraction are known.

Determination of v_v requires some discussion. It can be estimated by cutting sections through the material and optically measuring the fraction of area occupied by voids. There are instruments available to do this. Another approach involves direct measurement of composite density. Equation (1.2-5) can be written

$$v_v = 1 - v_f - v_m$$

Using Equations (1.2-15) and (1.2-16) the following expression for void content is obtained:

$$v_v = 1 - \frac{\varrho_c}{\varrho_f} w_f - \frac{\varrho_c}{\varrho_m} w_m \qquad (1.2\text{-}19)$$

The expressions derived above are for average values of volume fractions and weight fractions. It should be noted that because fibers are never distributed uniformly throughout the material, local values of v_f and w_f can differ substantially from the averages.

Elastic Properties

Although composites are heterogeneous on the microscopic level, for design purposes they are usually considered to be homogeneous, anisotropic materials. The fact that the material is composed of fibers and a matrix with different properties is neglected, and it is simply treated as a material whose properties vary with direction.

Because the elastic characteristics of composites are more complex than those of isotropic materials, it takes more constants to describe them. Isotropic materials, such as most metals, can be characterized by two independent elastic constants. The most common sets of

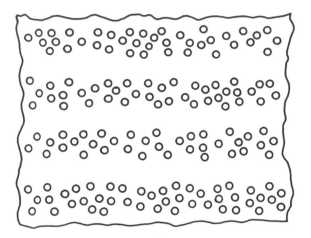

FIGURE 1.2-3.

elastic constants are Young's modulus, E, and Poisson's ratio, ν, or Young's modulus and shear modulus, frequently denoted by G. Only two of these three elastic constants are independent. If two are known, the third can be determined from the relationship [2]:

$$G = \frac{E}{2(1 + \nu)} \qquad \text{Isotropic} \qquad (1.2\text{-}20)$$

An isotropic material has the same elastic constants in every direction; however, for anisotropic materials, four or more constants are required [3].

Consider a unidirectional composite which, it will be recalled, is reinforced with continuous, straight, parallel fibers—as shown in Figure 1.2-1. A coordinate system is chosen so that the 1-axis is parallel to the fiber directions. For thin materials, the 2-axis is usually chosen to lie in the plane of the material, and the 3-axis denotes the through-the-thickness direction.

Twenty-one independent elastic constants are required to describe the most general anisotropic materials [3]. Fortunately, the situation is not so bad for fibrous composites. The number of constants required to describe an anisotropic material depends on the amount of symmetry it possesses. When the material on one side of a plane is the mirror image of that on the other side, the plane is referred to as a plane of symmetry of the material. In a composite, the symmetry of the material depends on the geometric arrangement of fibers. The 2-3 plane is a plane of symmetry for a unidirectional composite [3].

When a material has two orthogonal planes of symmetry, then the plane perpendicular to those two planes is also a plane of material symmetry. Such a material is called *orthotropic*. As a rule, the most complex unidirectional material can be considered to be orthotropic with the 1-2 and 1-3 planes being planes of material symmetry.

The extensional modulus in the fiber (longitudinal) direction is not very sensitive to the geometric arrangement of fibers in the material. However, the modulus in the transverse direction and the shear moduli in the 1-2 and 1-3 planes are strongly affected by the way in which fibers are distributed. In some unidirectional composites, such as those made from preimpregnated material, fibers may be grouped in layers, with resin-rich regions between them. Figure 1.2-3 illustrates this situation. If this occurs, the extensional moduli in the 2- and 3- directions will differ, as will the shear moduli in the 1-2 and 2-3 planes. For practical purposes, a material of the kind shown in Figure 1.2-3 can be considered to be orthotropic. Note that in a real material, the fiber locations are never perfect reflections about 1-2 and 1-3 planes. However, in a statistical sense, the material can be considered orthotropic.

An orthotropic material has nine independent elastic constants. There are several types of elastic constants used for anisotropic materials, such as stiffnesses and compliances, which are discussed in Volume 2.

A convenient set of elastic constants which can be measured directly and have physical significance are the three extensional (Young's) moduli, E_1, E_2 and E_3; the three shear moduli, G_{23}, G_{31} and G_{12}; and the three Poisson's ratios, ν_{23}, ν_{32} and ν_{12} [3]. The quantity E_1 is the extensional modulus in the fiber direction, and E_2 and E_3 are the moduli along the 2-axis and 3-axis, respectively. G_{12} refers to the shear modulus in the 1-2 plane, and G_{23} and G_{31} have similar meanings. The Poisson's ratio ν_{12} is defined as the ratio of the compressive strain along the 2-axis to the tensile strain in the 1-direction when a tensile load is applied in the 1-direction. This is the most common definition of ν_{12}, but the opposite one occasionally appears in the literature. Note that for composite materials, generally ν_{12} is not equal to ν_{21}. However, the two are related by the expression [3]:

$$\frac{\nu_{12}}{\nu_{21}} = \frac{E_1}{E_2} \qquad (1.2\text{-}21)$$

Similar expressions hold for ν_{23} and ν_{31}.

For most unidirectional materials, E_1 is greater than E_2 by an order of magnitude or more. Inspection of Equation (1.2-21) shows that this statement implies that the Poisson's ratio ν_{12} is much greater than ν_{21}. This fact is confirmed experimentally [3]. The physical explanation for this difference is that because the stiffness in the fiber direction is very great, there is considerable resistance to contraction along the 1-axis when the material is pulled in the lateral (2) direction. However, since the transverse stiffness E_2 is much smaller than E_1, when the composite is pulled in the axial (1) direction, there is much less resistance to transverse contraction.

For similar reasons, ν_{31} is usually much smaller than ν_{13}. However, since the transverse stiffnesses E_2 and E_3 are usually the same order of magnitude, the differences between ν_{23} and ν_{32} are usually much smaller than those between ν_{21} and ν_{12}, or ν_{31} and ν_{13}.

The relationship between the elastic moduli of a composite and those of its constituents, fibers and matrix is a complex topic. However, there is a simple expression for axial modulus that is very useful in evaluating the validity of experimental data. To a first approximation, regardless of how the fibers are arranged internally, the extensional modulus in the fiber direction is given by [3]:

$$E_1 = v_f E_f + v_m E_m \qquad (1.2\text{-}22)$$

where E_f is the fiber axial Young's modulus, E_m is the matrix modulus and v_f and v_m are the fiber and matrix volume fractions discussed earlier. Figure 1.2-4 illustrates this relationship. An equation of this form, in which a property of the composite is equal to the sum of fiber and matrix properties weighted by volume fraction, is referred to as the "rule of mixtures." It should be emphasized that the rule of mixtures is not valid for most composite properties. Longitudinal extensional modulus is one of the exceptions. Furthermore, the transverse moduli E_2 and E_3 and the shear moduli are sensitive to fiber geometry [4–11]. Figure 1.2-5, redrawn from the work of Adams and Tsai [12] illustrates both the nonlinear relationship between transverse modulus and fiber volume fraction, and the dependence on fiber geometry for glass/epoxy composites. The curve labeled "square array" represents the predicted modulus for this fiber geometry based on the analysis of Adams and Doner [13]. Also shown are the analytical results for a hexagonal array obtained by

Foye [14] and the experimental data of Adams, Doner and Thomas [15].

It is very common to find published data for E_1 that do not satisfy Equation (1.2-22). Possible sources of error are measurement of fiber volume fraction, modulus, or both; and differences between assumed and actual values of fiber and matrix modulus. With the exception of glass fibers, actual fiber moduli can differ substantially from the nominal values presented in manufacturers' literature.

When fibers are dispersed randomly throughout the composite, as shown in Figure 1.2-6, there are no preferred directions in the transverse plane, and on the average, the extensional modulus is the same in every direction in the plane. Such material is called transversely isotropic; regardless of the choice of directions for the 2- and 3-axes, E_2 and E_3 are equal, and ν_{23} and ν_{32} are equal.

If properties in the transverse plane are independent of direction (transverse isotropy), $\nu_{13} = \nu_{12}$ and $G_{31} = G_{23}$. Note, however, that $\nu_{12} \neq \nu_{21}$ and $\nu_{13} \neq \nu_{31}$. These quantities are still related by relations of the form of Equation (1.2-20). A transversely isotropic material has five independent elastic constants [3]. A convenient set is E_1, E_2, G_{12}, ν_{13} and G_{23} or ν_{23}.

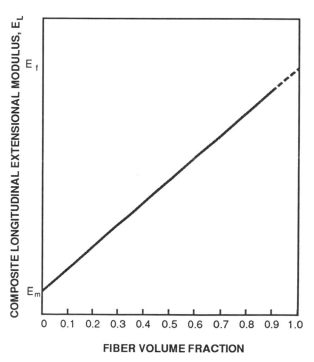

FIGURE 1.2-4.

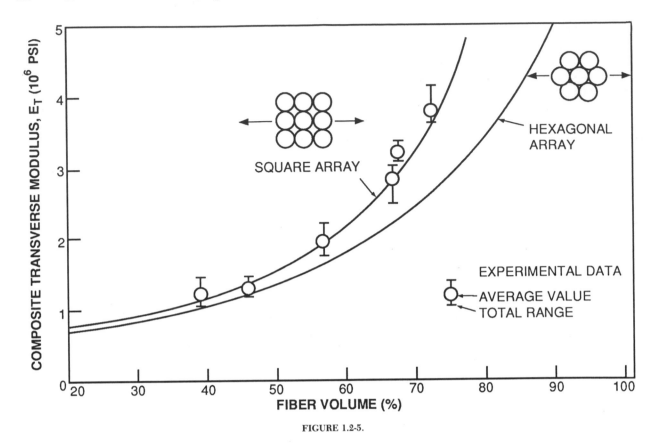

FIGURE 1.2-5.

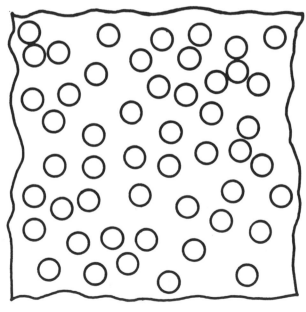

FIGURE 1.2-6.

Because of isotropy in the transverse plane, E_2, ν_{23} and G_{23} are related by the expression

$$G_{23} = \frac{E_2}{2(1 + \nu_{23})} \quad \text{Transverse Isotropy} \quad (1.2\text{-}23)$$

This expression has the same form as Equation (1.2-20) which relates shear modulus, Young's modulus and Poisson's ratio for an isotropic material.

A great many applications have components in the form of thin plates or shells. Some examples are automobile drive shafts, door panels and hoods, aircraft and wing skins, pressure vessels, and chemical storage tanks. For such structures, it is common to assume that the components are in a state of plane stress; that is, deformations and stresses through the thickness are neglected. As a result, the material constants in the plane of the structure are of primary interest. (An exception to this simplification is stress discontinuities, such as in joints where through-the-thickness properties are important.)

For the case of plane stress, there are four independent elastic constants of interest: a convenient set is E_1, E_2, G_{12}, and ν_{12} [3]. This simplification holds for both orthotropic and transversely isotropic materials. These four elastic constants are the ones that are commonly evaluated for most unidirectional composites, and the ones that are tabulated in a subsequent discussion. It will be recalled that E_1 is the extensional modulus in the fiber direction, E_2 is the transverse extensional modulus, and G_{12} is the axial shear modulus which defines the shear deformation resulting from shear stress parallel to the 1-2 axes. The Poisson's ratio ν_{12} is defined as the ratio of the transverse compressive strain to the axial tensile strain when a tensile stress is applied in the axial direction [3].

For thin unidirectional materials, it is common to find another set of symbols used to represent the four elastic constants discussed above. The fiber or longitudinal axis is designated by the subscript L and the transverse by T. Using this convention, the elastic constants are

$$E_L = E_1 = \text{axial (longitudinal) extensional modulus}$$
$$E_T = E_2 = \text{transverse extensional modulus}$$
$$G_{LT} = G_{12} = \text{axial shear modulus}$$
$$\nu_{LT} = \nu_{12} = \text{axial Poisson's ratio}$$

These are the basic material elastic properties that are used by a designer.

Strength Properties

The strength properties of unidirectional composites are governed by complex internal failure modes and cannot be described as simply as elastic properties. For example, when the five elastic constants of a transversely isotropic material are known, the elastic response under any arbitrary loading condition is known exactly. The same is not true for strength. At present, there is no universally accepted law for the prediction of composite failure under arbitrary loading conditions. This subject will be discussed in more detail in Volume 2. Although there is no general agreement on how to predict composite failure under combined loads, there are a number of basic strength properties. For a unidirectional material they are longitudinal (axial) and

transverse tensile strength; axial and transverse compressive strength; and axial, in-plane, and interlaminar shear strengths [3]. The interlaminar shear strength is the shear strength through the thickness of the material in a plane containing the fiber axis. It gets its name from the fact that unidirectional composites are frequently made by stacking up thin layers (laminas). Interlaminar shear strength is frequently called "short beam shear strength," after the test by which it is commonly evaluated [16]. This procedure is described in Test Methods, Volume 6.

There are many symbols used to denote the strength properties listed above. The system used in the U.S. Air Force Advanced Composites Design Guide, which is widely used, is shown below. An alternative notation (the X_i^j) which appears in some strength theories is also shown where applicable.

$$X_1^T = \begin{cases} F_L^{ty} = \text{axial (longitudinal) tensile "yield" stress} \\ F_L^{tu} = \text{axial (longitudinal) tensile ultimate stress} \end{cases}$$
$$X_1^C = \begin{cases} F_L^{cy} = \text{axial (longitudinal) compressive "yield" stress} \\ F_L^{cu} = \text{axial (longitudinal) compressive ultimate stress} \end{cases}$$
$$X_2^T = \begin{cases} F_T^{ty} = \text{transverse tensile "yield" stress} \\ F_T^{tu} = \text{transverse tensile ultimate stress} \end{cases}$$
$$X_2^C = \begin{cases} F_T^{cy} = \text{transverse compressive "yield" stress} \\ F_T^{cu} = \text{transverse compressive ultimate stress} \end{cases}$$
$$S_6 = \begin{cases} F_{LT}^{sy} = \text{in-plane shear "yield" stress} \\ F_{LT}^{su} = \text{in-plane shear ultimate stress} \end{cases}$$
$$F^{isu} = \text{interlaminar shear ultimate stress}$$

The corresponding strains are denoted by ϵ, with the appropriate subscripts and superscripts.

Note that the composites treated in this work do not exhibit a true plastic behavior as metals do. However, in some cases their stress-strain curves display some nonlinearity that appears similar to that of metals. There are many sources of nonlinear behavior, such as resin crazing (formation of small cracks); fiber buckling in compression; debonding of fibers; viscoelastic deformation of the matrix, fibers, or both; etc. The term *"yield" stress* is used as a convenience to denote departure from linearity. The designer should take note of any departure from linearity in material behavior, as it may imply irreversible internal damage that can affect

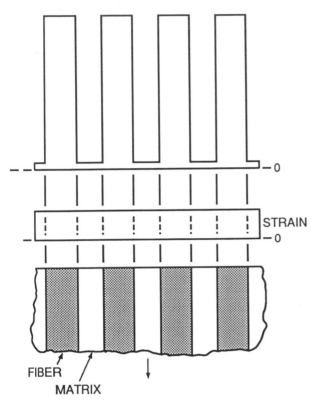

STRAIN

FIBER

MATRIX

FIGURE 1.2-7.

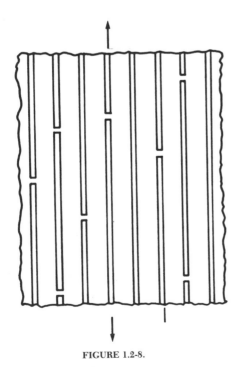

FIGURE 1.2-8.

performance. On the other hand, there is no such thing as a perfectly linear material, so some judgment is required in assessing the importance of the observed nonlinearity. The significance of the yield can be evaluated in several ways: one is by direct visual observation of the material for signs of crazing or other internal damage. The most reliable way is to subject the material to fatigue and creep loading below and above the "yield" point to determine whether there is any significant change in behavior. These tests should include any environmental influences that could affect behavior, such as humidity, temperature, and chemical attack.

The axial tensile and compressive strengths are dominated by fiber properties, although matrix characteristics have an influence. The other strength values, which are often lumped into the general category of "transverse strength properties," are influenced primarily by matrix strength characteristics, fiber-matrix interfacial bond strength, and internal stress concentrations arising from factors such as voids and the proximity of fibers. As both axial and transverse strength properties influence design and performance, the modes of failure and the factors affecting these strength values will be considered.

Axial Tensile Strength

Most of the unidirectional composites of interest have high axial tensile strengths—one of the major reasons for their use. Composite tensile strength is governed by a number of factors: fiber modulus and strength properties; resin mechanical properties including stress-strain behavior, viscoelastic properties and strength characteristics; and fiber-matrix interfacial strength characteristics.

Consider a long unidirectional composite specimen loaded in tension. The internal state of stress at the ends of the composite is very complex and depends on the manner of load introduction. In the central portion of the specimen, the axial strain is uniform. Although the axial strain in the fibers and matrix is the same, the axial stress in the fibers is much greater than that in the matrix because fiber modulus is at least an order of magnitude greater for most composites. (See Figure 1.2-7.) If the Poisson's ratios of the fiber and matrix are different, the two materials will tend to contract by different amounts laterally to induce a complex transverse stress distribution. This complication is generally of secondary importance and will not be considered here.

The total axial loads carried by the fibers and matrix phases of the material are equal to the products of their respective stresses and cross-sectional areas. For most practical unidirectional composites, the fiber cross section is at least as large as that of the matrix. As a result, the fibers carry most of the applied load, and the direct contribution of the matrix is small; however, the matrix does have an important indirect influence on tensile strength, as we shall see.

Consider what happens as the material is loaded to failure. We noted in subsection 1.1.2 that the reinforcing fibers of interest do not have a unique tensile strength. The strengths of individual fibers vary widely. Therefore, at a low stress level, one of the fibers in the composite may break. When it does so, it releases energy which may cause a propagating crack or stress wave that initiates an immediate and catastrophic mode of failure [17]. Fortunately, this break does not occur in most resin-matrix composites, although there is evidence that it may occur in composites combining boron fibers and an aluminum matrix [18].

When a first fiber fracture does not result in composite failure, additional load can be applied. As the stress level increases, additional fibers break randomly throughout the material. These fiber breaks have been observed directly in a special glass/epoxy material [19] and indirectly using acoustic emission detectors [20]. Figure 1.2-8 illustrates this process.

Each of the scattered fiber breaks causes a local perturbation of stress that depends on fiber modulus, matrix stress-strain properties and strength, and interfacial bond strength. When a fiber breaks, the broken ends tend to separate. Assuming perfect bonding, the matrix resists this displacement through shear stress on the lateral surface of the fiber. Figure 1.2-9 illustrates this situation. The shear stress induces load into the fiber so that while the axial tensile stress in the fiber is zero at the point of the break, the fiber is fully loaded at some distance along its length. Figure 1.2-10 presents a schematic representation of the interfacial shear stress and fiber tensile stress near a break. The distance along the fiber required to reload the fiber to its fully loaded stress level is frequently called the transfer length ℓ_T. This quantity is referred to as the ineffective length, δ, in reference [19]. The total distance over which the stress is perturbed is twice this length. In references

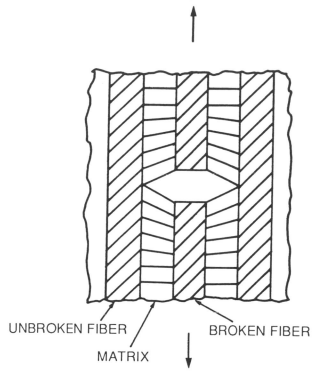

UNBROKEN FIBER BROKEN FIBER

MATRIX

FIGURE 1.2-9.

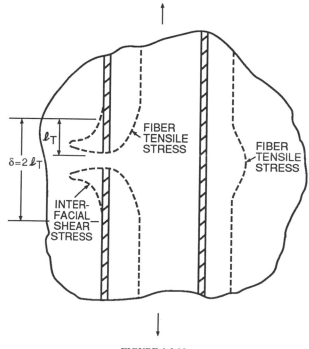

FIBER TENSILE STRESS

FIBER TENSILE STRESS

ℓ_T

$\delta = 2\ell_T$

INTER-FACIAL SHEAR STRESS

FIGURE 1.2-10.

[17,21] ineffective length is used to denote this quantity. In this discussion we use the latter definition because it has greater physical significance for composite tensile strength.

Ineffective length is an important characteristic dimension of a composite material for tensile strength because fiber mean tensile strength depends on length, as discussed in detail in subsection 1.1.2. The relevant fiber tensile strength associated with composite tensile strength is the mean strength of fibers of length δ. This concept is important. The plots of fiber mean strength as a function of length presented earlier showed that the projected mean strengths of fibers as long as those used in large structures are very low. Fortunately, composite tensile strengths are significantly greater than these values. The reason is that the matrix localizes the effect of fiber breaks; thus, ineffective length, rather than overall structure length, becomes the reference dimension for tensile strength. However, some data suggest that composite strength may decrease as composite volume increases. The effect, if it exists, is far less severe than that seen in fibers.

Ineffective length depends on the mechanics of stress transfer between the fiber and matrix. Friedman [21] derived an approximate expression for the case of a perfectly bonded elastic fiber in an elastic matrix:

$$\delta = d_f \left[\left(\frac{1}{2} \right) \left(\frac{E_f}{G_m} \right) \left(\frac{1 - V_f^{1/2}}{V_f^{1/2}} \right) \right]^{1/2} \qquad (1.2\text{-}24)$$

where

d_f = fiber diameter
E_f = fiber extensional modulus
G_m = matrix shear modulus
V_f = fiber volume fraction

For most materials, the values of ineffective length computed using Equation (1.2-24) are less than ten fiber diameters. It will be recalled that the diameters of most fibers are 13×10^{-6} m (5×10^{-4} in) or less. Therefore, the elastic ineffective length for most composites is of the order of 10^{-4} m (10^{-3} in).

As Figure 1.2-10 shows, there is a local shear stress concentration at the end of the broken fiber which can cause localized failure of the matrix or interfacial bond.

When this failure occurs, the elastic analysis on which Equation (1.2-24) is based is no longer valid. Several other methods have been proposed to treat this case [22–24]. These analyses show that the ineffective lengths associated with matrix or interface failure are larger than those of the elastic case. Nevertheless, they are generally still quite small, say of the order of 10^{-3} m (10^{-2} in), and the fiber mean strengths associated with these lengths are relatively high.

So far, we have discussed the role that the matrix plays in localizing the effects of fiber breaks. Due to this effect composites do not display the severe reduction in strength with increasing length found in fibers. There are other aspects of the stress perturbation near fiber breaks that influence composite tensile strength. The shear stress in the matrix transfers load from the broken fiber to the unbroken fibers in the vicinity which causes local stress concentrations in these surrounding fibers. Figure 1.2-10 shows an idealized representation of this stress. The magnitude of these stress concentrations depends on the number of fibers in the vicinity of the break. For example, using an approximate model, Hedgepeth [25] found that for a two-dimensional array of elastic fibers perfectly bonded to an elastic matrix —that is, a single layer of fibers—the average tensile stress in the two intact fibers next to a broken one is one-third greater than the average fiber stress far from the break. That is, the stress concentration factor is 4/3. When the fibers in a unidirectional composite are arranged in a hexagonal array, the stress concentration factor for each of the six fibers surrounding a broken one is 1.10 [26].

As discussed in subsection 1.1.2, the strength of a fiber in any region along its length is a statistical variable—and not a unique quantity. The effect of the stress concentrations is to increase the probability of failure of the overstressed fibers in the region of the fiber break. This problem has been treated in references [17] and [27–30]. When an overstressed fiber breaks, the stress concentration in the surrounding fibers increases. For example, the stress concentration for two adjacent broken fibers in a two-dimensional array is 1.6 compared to 4/3 for one broken fiber. The increase in stress intensity that occurs when an overstressed fiber breaks results in a higher probability of failure of the surrounding fibers. This process can continue, giving rise to a mode of failure associated with propagating fiber breaks.

From the above discussions, we see that the internal processes that give rise to tensile failure are complex, and several possible failure modes have been identified. There are other factors that can influence failure. For example, local matrix or interface failure tends to reduce the intensity of fiber stress concentrations [23,24]. In addition, when interfacial bond strength is high, fiber fracture can cause penny-shaped cracks in the matrix [31].

There is one more topic related to composite tensile strength that deserves consideration—size effect. It is well-established that the strength of flaw-sensitive materials like brittle ceramics and glasses decreases with increasing volume because the probability of finding a serious defect increases with increasing volume [32]. Size (length) effect, as seen in glass, graphite and aramid fibers, was discussed in subsection 1.1.2.

In a classic work on the subject Weibull [32] proposed an analytical method to treat the problem of flaw-sensitive brittle materials. It is not clear whether composites can be placed in this class of materials. However, the study of tensile failure mechanisms led to the prediction of a possible size effect in composites [17].

There is some experimental evidence of a size effect in composites [33], but it is far from conclusive. The strength of a specimen loaded in flexure is usually greater than that of the same specimen loaded in tension. In reference [34], the strength difference was attributed to the fact that a smaller volume of material is subjected to tensile stress in a flexure test and the use of Weibull's theory gave good correlation with data for graphite/epoxy. This approach was also used in the analysis of aramid/epoxy flexural behavior with good results [35]. However, it may be that other explanations can explain the difference in failure levels, such as the existence of a stress gradient in the flexure test. A tendency toward lower strengths with increasing volume was reported in work on glass fiber-reinforced rocket motor cases [36]. However, the data are clouded by the fact that the state of stress in a large pressure vessel, such as a rocket case, is extremely complex and is far from pure tension.

For the designer, the significance of all this analysis is that some caution is in order in using stress data obtained from small laboratory coupons for structures which may be many orders of magnitude larger. Although this discussion has focused on the possible size

effect in tensile strength, a similar argument may apply to other composite strength properties.

Axial Compressive Stress

When unidirectional composites are subjected to axial compression, the fibers support most of the applied load—as they do in the case of axial tension—because of the much greater longitudinal stiffness of the fibers. Therefore, the axial stress distribution pictured in Figure 1.2-7 for tensile stress also applies to compression loading.

When isolated fibers are loaded in compression, they buckle at very low stress levels unless their length is extremely small. Despite this buckling, the compressive strength of unidirectional composites, such as the compressive strength of some graphite/epoxy and boron/epoxy systems, can be quite high. The explanation for this apparent anomaly is that the matrix stabilizes the fibers, preventing them from buckling at low stress levels.

As in the case of axial tensile stress, much of what we know about the mechanisms of failure associated with axial compression comes from observation of failed specimens, from tests performed on specially designed model materials, and from analytical treatments of the problem. The analyses are usually based on experimental evidence and intuitive models of material behavior arising from fundamental principles of mechanics. Some possible failure mechanisms that have been identified are fiber compressive strength failure, various forms of fiber and material internal instability (buckling) and fiber debonding or matrix fracture caused by local shear stress. It is important to note that the type of instability discussed here is not overall buckling of the material component or specimen as a structure. Rather, we are talking about internal instabilities within the material itself. In fact, when compression strength is measured using coupons, great care is taken to assure that the specimen does not fail by buckling as a Euler column. This subject will be discussed in greater detail in Volume 6 (Test Methods).

It is frequently difficult to determine whether the failure of a composite loaded in axial compression was initiated by fiber compressive failure or instability. However, it is well-established that the compressive

MICROINSTABILITY FAILURE MODES
(FIBER BUCKLING)

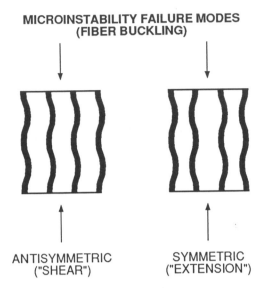

ANTISYMMETRIC   SYMMETRIC
("SHEAR")   ("EXTENSION")

FIGURE 1.2-11.

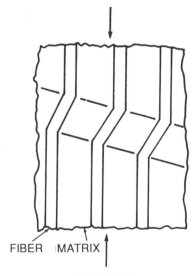

FIBER   MATRIX

FIGURE 1.2-12.

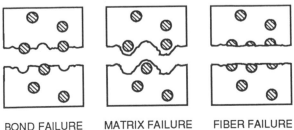

BOND FAILURE   MATRIX FAILURE   FIBER FAILURE

FIGURE 1.2-13.

strength of unidirectional composites reinforced with Kevlar 49 aramid fibers is limited by the failure of the fibers themselves [37,38]. Kevlar 49 fibers do not fail catastrophically under compressive loading. Instead, the mode of fiber failure appears to be one of kinking. The fiber appears to behave like a "microcomposite" that fails by a mode of internal instability [39].

A number of authors have proposed models for fiber instability. For example, simple, two-dimensional models, shown in Figure 1.2-11, were proposed at about the same time by Schnerch [40] and Rosen [41]. Another mode of instability, kinking, is shown schematically in Figure 1.2-12 [42]. As is generally the case with buckling phenomena, fiber instability is sensitive to imperfections such as local curvature and misalignment. When compressive load is applied to curved fibers, shear stress develops between the fiber and matrix which can result in debonding or matrix cracking, causing overall material failure [43].

Transverse Strength Properties

As discussed above, strength properties other than axial tension and compression are sometimes lumped together under the heading of transverse strength properties. They include transverse tensile and compressive strengths and axial and interlaminar shear strengths. These quantities are very dependent on matrix strength properties, interfacial properties and imperfections such as voids and microcracks resulting from resin cure shrinkage and from internal thermal stresses. These thermal stresses may be induced by large temperature changes such as those associated with cool down after elevated temperature cure. Generally speaking, factors that affect one transverse strength property affect them all, with the possible exception of transverse compressive strength, which is less sensitive to imperfections.

Figure 1.2-13 illustrates schematically some of the internal failure mechanisms and associated factors affecting transverse strength properties. When the fiber-matrix bond strength is low, debonding occurs and the failure surface includes many debonded resin surfaces. For the case where bond strength and fiber transverse strength are high, the failure surface tends to run primarily through the matrix. When fiber transverse strength is low and matrix and bond strength are high, transverse fiber failure occurs. This mode of failure appears to be a limiting factor in the transverse strength properties of aramid fiber composites.

The significance of resin strength on transverse strength properties is illustrated in Figure 1.2-14, based on the work of Brelant and Petker [44], which shows how composite interlaminar shear strength varies with resin tensile strength for a glass fiber-epoxy composite. We see that interlaminar shear strength increases with increasing resin strength until the latter reaches a value of about 83 MPa (12×10^3 psi). Above this value, further increases in resin tensile strength do not result in any improvement in shear strength. The apparent reason is that the limiting factor has become fiber-matrix bond strength. Further increases in interlaminar shear strength can be obtained only by improving bond strength.

As discussed earlier, internal defects, such as voids,

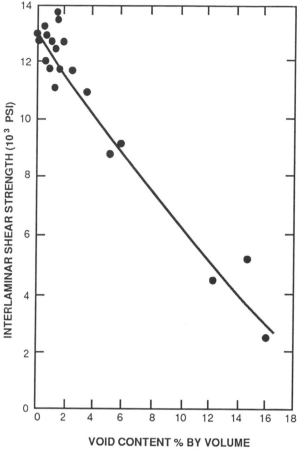

FIGURE 1.2-15.

can also have an important effect on transverse strength properties. Figure 1.2-15 shows how interlaminar shear strength varies with void content for glass/epoxy. The figure shows the severe drop in strength associated with high void content. The apparent reason for the sensitivity to void content is that voids cause severe internal stress concentrations in the material. Further results are provided in reference [45].

Internal stress concentrations exist even in materials with extremely low void contents. This is because the fibers and matrix have different transverse stiffness properties. For example, Adams and Doner [46] studied the internal stresses in a unidirectional composite with a square array of fibers subjected to axial shear. They found that the shear stress at some points in the material is significantly higher than the applied shear stress. The magnitude of the stress concentration factor (ratio of

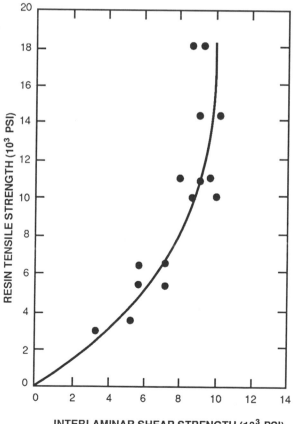

FIGURE 1.2-14.

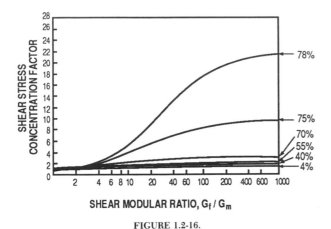

FIGURE 1.2-16.

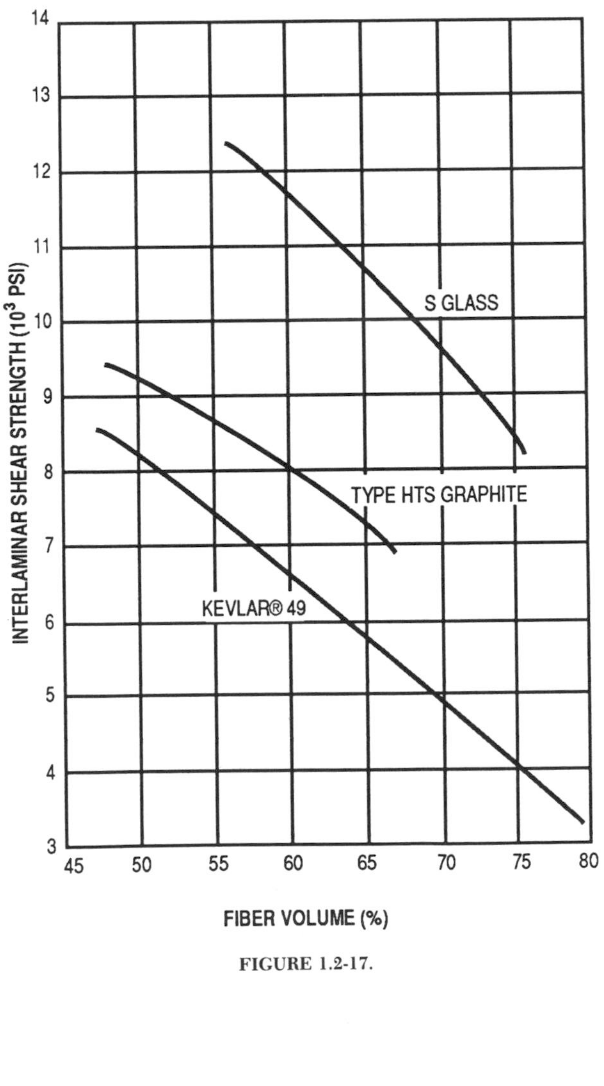

maximum stress-to-applied stress) increases as average fiber spacing decreases. This increase in the ratio occurs when the fiber volume fraction, v_f, increases. Adams and Doner also found that the stress concentration factor increases as the ratio of axial fiber shear modulus-to-matrix shear modulus, G_f/G_m, increases. For most composites the shear stress concentration factor is in the range of 2 to 3, as shown in Figure 1.2-16.

The predicted increase of shear stress concentration factor with increasing fiber content is in qualitative agreement with the experimental interlaminar shear strength data of Woodberry and Bormeier [47] for S glass, HTS graphite, and Kevlar 49 aramid composites, as shown in Figure 1.2-17.

We note that the dependence of shear strength on volume fraction is very strong which is consistent with the predictions of Adams and Doner [46]. However, the distribution of fibers in a real composite is greatly different from that of a regular, square array; therefore, we should not expect to find precise agreement between theory and experimental data. Nevertheless, analyses of this sort provide valuable insights into factors affecting composite strength. There is another reason that we should not expect agreement between predicted elastic stress concentration factors and observed strength: once local failure has occurred at some regions of the material, there is a complex redistribution of stress, and the elastic analysis is no longer valid.

In this discussion, we have emphasized the factors affecting shear strength because there is a considerable body of data on this property. The primary reason for this data accumulation is that the short beam shear test, which is commonly used to measure shear strength, is simple and inexpensive to perform. However, as mentioned above, the factors affecting shear strength have qualitatively similar effects on the other transverse strength properties—with the possible exception of transverse compression strength. For example, stress concentrations also arise in the case of transverse tensile stress. Figure 1.2-18 shows the shear and normal tensile stresses at the fiber-matrix interface for transverse tensile loaded composites with a square fiber array [15]. Stresses are normalized with respect to the applied tensile stress. The assumed ratio of fiber transverse extensional modulus-to-matrix modulus is 120, which is characteristic of boron/epoxy. Two different fiber volume contents are represented: 55% and 75%. Assumed fiber and matrix Poisson's ratios are 0.20 and

FIGURE 1.2-17.

0.35, respectively. We see that the maximum interfacial tensile stresses are two to three times as great as the applied stress. As for the case of axial shear, stress concentrations increase with increasing fiber content (decreasing spacing).

We noted earlier that internal thermal stresses resulting from cool down after elevated temperature cure can be quite large. The stresses result from the fact that the coefficient of thermal expansion of most resins is substantially greater than the radial coefficient of thermal expansion of glass, graphite and most other fibers. The main exceptions are aramid fibers, which tend to have large radial coefficients of thermal expansion.

Because of their greater coefficient of thermal expansion, resins tend to contract more than fibers, resulting in a compressive interfacial normal stress when composites are cooled. This problem was studied by Adams and Doner [13]. They investigated the influence of the ratio of fiber transverse stiffness-to-matrix stiffness on interfacial stress for square array fibers having a transverse coefficient of thermal expansion of $4.9 \times 10^{-6}/K$ $(2.7 \times 10^{-6}/°F)$ in a matrix with a coefficient of expansion of $45 \times 10^{-6}/K$ $(25 \times 10^{-6}/°F)$. Assumed values for fiber and matrix Poisson's ratios were 0.20 and 0.35, respectively. Fiber volume fraction was 0.70. Figure 1.2-19 shows how the maximum interfacial compressive stress varies with stiffness ratio for a temperature drop of 111 K (200°F). The maximum stress for glass/epoxy with a modulus ratio of about 20 is approximately 52 MPa $(7.5 \times 10^3 psi)$. The magnitude of the predicted thermal stresses arising from cool down is seen to be large in comparison to the transverse strength properties of the composite. It is possible that viscoelastic stress relaxation may reduce the magnitude of these stresses somewhat. However, it is not unusual to observe transverse cracking in some composite systems subjected to large temperature excursions.

1.2.3 Fabric Composites

Woven reinforcing fabrics are made by interlacing individual filaments, ends (untwisted fiber bundles), yarns (twisted fiber bundles), and rovings. Fabric composites have mechanical properties similar to those of laminates made from orthogonal unidirectional layers.

NORMALIZED STRESS

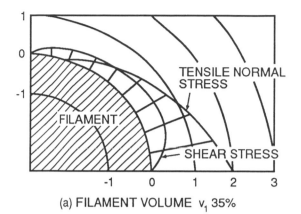

(a) FILAMENT VOLUME v_1 35%

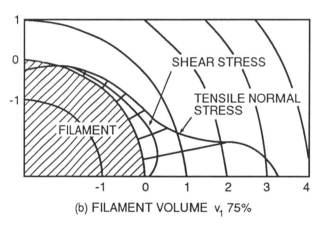

(b) FILAMENT VOLUME v_1 75%

FIGURE 1.2-18.

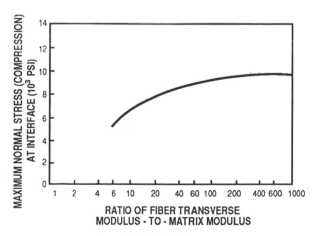

FIGURE 1.2-19.

FIGURE 1.2-20. Plain weave.

However, fiber curvature arising from yarn twist and weave crimp makes fabrics significantly less efficient reinforcements than aligned, straight fibers. Fabric composites often have lower fiber volume fractions than those made from unidirectional materials, which also contributes to lower elastic moduli and strength properties. Fiber crimp causes local stress perturbations which result in lower tensile and compressive strengths.

The major fabric styles are plain weaves, twills, satins and woven rovings. Fabric construction determines reinforcing efficiency, conformability to complex surfaces (drape) and resistance to distortion (stability). Plain weaves are relatively inefficient, have poor drape, but are very stable. Long-shaft satins are among the most efficient reinforcing fabrics, and they conform well to complex surfaces.

Fabric processing and handling frequently results in a type of distortion called skewing in which yarns are no longer orthogonal. Composites made from skewed fabrics tend to warp when subjected to a mechanical load or temperature change.

Fabrics are materials which are formed from fibers by a variety of processes such as weaving, knitting and felting. In this subsection we concentrate on woven fabrics which are the major type used to reinforce composites. Woven fabrics are produced by interlacing individual filaments, untwisted fiber bundles (ends), twisted fiber bundles (yarns), and bundles of ends (rovings). There are many fabric styles used as reinforcements. We consider some of the major forms.

With the exception of some specialty fabrics, which we will touch upon briefly later on, woven reinforcements consist of orthogonal fibers. The long direction of the fabric is called the warp, and the width direction is variously referred to as the fill, woof or weft. Fill yarns are often referred to as picks. (In this discussion we use the term *yarn* to denote the basic weaving material with the understanding that it represents single fibers, ends, or rovings.) When a fabric contains the same number of equal weight warp and fill yarns per unit length, and the weaving patterns in these directions are the same, the fabric is said to be balanced—if not, it is unbalanced. When the amount of yarn per unit area in one direction, usually the warp, is much greater than the other, say at least 80/20, the fabric is said to be unidirectional. Unidirectional fabrics should not be confused with unidirectional composites which are reinforced with unwoven, parallel, straight fibers.

Fabric construction is described by its pattern and its warp and fill yarn type, weight and spacing. Yarns are typically characterized by their mass per unit length, in units such as denier, tex or kg/m. Spacing is usually expressed as yarns per unit length. Many fabrics are woven from groups of yarns that are individually twisted and then twisted together.

The simplest pattern is the plain weave in which each warp and fill yarn passes over one yarn and under the next, as in Figure 1.2-20. Plain weave fabrics are the most stable of all woven materials with orthogonal yarns. By stable, we mean they resist distortion. However, they are generally not as efficient structurally as satin weaves and twills, and they have poor drape characteristics, that is, they do not conform easily to surfaces with double curvature.

In twill weaves, one or more warp yarns pass (float) over at least two consecutive fill yarns. These fabrics have characteristic diagonal patterns known as twill lines (Figure 1.2-21). Twills are relatively stable and more structurally efficient than square weaves, and they have relatively good drape.

The other main reinforcing fabric patterns are the

satins: a single warp yarn passes over three or more picks and then under one. In the crowfoot satin weave, the float length is three yarns. When the float is greater than three yarns, the pattern is frequently called a long shaft satin. Figure 1.2-22 shows a five-shaft satin, alternately called a five-harness satin. In this pattern, a yarn passes over four others and under one. Long-shaft satins are the most efficient reinforcing fabrics. Because the yarns have long floats and are unconstrained along much of their length, they conform well to surfaces with compound curvature. The basket weave is a variant of the plain weave in which two or more yarns are grouped together, and the groups alternately pass over and under one another.

Woven rovings are heavy fabrics made from rovings, as the name implies. They are widely used in boat construction, cargo containers, and other applications where high performance materials are not needed and thick laminates are required. Many woven rovings are not balanced, having more warp than fill rovings per unit length to allow higher loom production rates than are possible with balanced fabrics. A typical warp-to-fill ratio is 5/4. As a result, properties of woven roving laminates are significantly higher in the warp direction than in the fill. Boat hulls frequently are made by alternating plies of woven roving and mat, since it is believed that these laminates have higher interlaminar shear strengths than those made entirely of woven rovings. Because of their coarseness, the surface of woven roving laminates is rough. Where appearance is important, mat or finer fabrics are used on the surface.

Very light, open weave fabrics made from fine yarns tend to be "sleazy." That is, they distort easily. To obtain stability, the leno weave is often used. In this type of construction, two or more warp yarns cross over each other, locking fill yarns in place. In the mock leno weave, which resembles the leno, stability is achieved without having the warp yarns cross one another.

There are a number of specialized fabrics that will be mentioned briefly. A few fabrics with non-orthogonal fiber orientations have been developed. One of these is "Doweave," named after its originator, Norris F. Dow. Among the major advantages of this construction are that it is very stable and resists tearing. Some patterns produce laminates that are elastically isotropic in the plane. Figure 1.2-23 illustrates a typical pattern.

To minimize the property reduction arising from crimp, weave patterns have been developed which have

FIGURE 1.2-21. Twill weave.

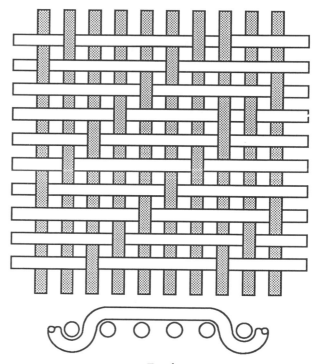

FIGURE 1.2-22. Five-harness satin.

FIGURE 1.2-23. Triaxial weave ("Doweave").

a high percentage of straight yarns held in place by lighter, crimped yarns. Several variations of these materials, sometimes called "high-modulus fabrics," are available with straight yarns in the warp direction, the fill direction, or both.

So far, all of the fabrics we have discussed have been planar, or two-dimensional. A variety of reinforcing materials have been developed having nonplanar fiber patterns that are generally referred to as three-dimensional fabrics [48,49]. These materials have been used in rocket motor nozzles, reentry vehicles, and aircraft brakes. For these applications, the material system frequently consists of graphite (carbon) fibers in a carbon matrix (commonly called carbon/carbon composites).

There are a number of reinforcing materials that are referred to as non-woven fabrics. Although they have radically different constructions and properties, their common feature is that they are made by bonding fibers together in a structured pattern, random pattern, or both, without any interlacing. From a mechanics standpoint, these materials can usually be regarded as combinations of unidirectional and random fiber reinforcements; they do not fall under the category of fabrics.

Mechanics of Fabric Reinforcement

We now examine reinforcement principles for fabric composites. We limit our consideration to orthogonal fabrics, although much of the discussion also applies to other types of woven materials. We consider several aspects of the mechanics of fabric reinforcement: macroscopic properties, reinforcement efficiency, internal deformations and failure mechanisms, and the effect of fabric distortion (skewing).

On the macroscopic level, the mechanical properties of balanced fabric composites are similar to those of laminates having equal numbers of unidirectional layers oriented at 90° to one another. Two examples of laminates with orthogonal unidirectional layers are [0/90] and [±45]. The first has fibers oriented along and perpendicular to the reference axis. The second has fibers at +45° and −45° to the reference axis. A fabric laminate with properties similar to those of a [0/90] composite can be obtained by orienting the fabric warp fibers parallel to the reference axis or perpendicular to it. By rotating the warp 45° to the reference axis, an analogy to the [±45] laminate is obtained. Unbalanced fabrics are analogous to laminates made from different numbers of layers in the two orthogonal directions.

As in the case of unidirectional and discontinuous fiber composites, the properties of fabric laminates depend strongly on fiber volume fraction. As a rule, fabrics do not pack as densely as unidirectional fibers. The result is that fabric composites tend to have lower fiber volume fractions than unidirectional composite laminates which is not true for all fabrics; laminates reinforced with long-shaft satins have been made with fiber volume fractions greater than 70%.

As we indicated in subsection 1.2.2, continuous, straight fibers provide the most efficient reinforcement. Most fabrics are made from continuous-fiber yarn. (Since asbestos is only available as discontinuous fibers, fabrics of this material are woven from staple yarns, which are made by twisting short fibers together.) However, fibers used in woven materials have some curvature over at least part of their length. Therefore, fabric composites are inherently less efficient as reinforcements than those made from unidirectional materials.

There are two main forms of fiber curvature in fabrics: twist and crimp. Many fabrics are made from yarns which are twisted because they are easier to weave and are less susceptible to damage than untwisted yarns. Crimp is curvature arising when fibers pass over and under one another.

Fabric reinforcing efficiency depends strongly on the percentage of fiber length that is curved and the severity of the curvature. Long-shaft satins tend to be relatively efficient because fibers are straight over much of their length. The influence of fabric construction on flexural strength and modulus was studied by Shibata, Nishimura and Norita [50] who considered 8-harness satin (8-shaft satin) and crowfoot satin fabrics having 10 ends (warp yarns) per cm with several different fill yarn spacings. The 8-harness satin fabrics all had 3,000-filament yarns. Crowfoot satins were made with both 3,000- and 1,000-filament yarns. One plain weave was also considered, but data for this fabric appear to be inconsistent and are omitted from consideration. We note that special problems encountered in testing fabrics may explain the apparent discrepancy [51].

Figure 1.2-24 shows how modulus varies with fill yarn spacing for epoxy matrix composites made from these fibers. Note that the fabrics with 10 picks per cm are balanced. Shibata et al. [50] normalized the data to 100% warp yarn fiber volume fraction. Fabrics made from 3,000- and from 1,000-filament yarns are marked 3k and 1k, respectively. The effect of increasing the number of fill yarns per cm is to increase fiber curvature. Consequently, it is not surprising to observe a general downward trend in modulus with increasing pick density. We note that the 8-harness satin modulus' insensitivity to pick spacing over the range tested can be explained by the fact that yarns in these fabrics are straight over most of their length. Consequently, the local increase in curvature at crossover points has a much smaller influence than for the crowfoot satins, where the float is only three yarns long.

The influence of crimp is illustrated in another way in Figure 1.2-24. We note that the crowfoot satins made from 1,000-filament yarns have consistently higher moduli than 3,000-filament yarns at the same spacing because crimp is more severe in fabrics made from heavier yarns for equal yarn spacing.

The variation in flexural strength with pick spacing is shown in Figure 1.2-25. We note that flexural strength is much more sensitive to fill spacing than is modulus because material strength characteristics are generally sensitive to local defects whereas moduli are volume average bulk properties. As for modulus, 8-harness satins are more efficient than the crowfoots, and the 1k crowfoot is more efficient than the 3k.

This study illustrates the important influence of crimp on the relative efficiencies of reinforcing fabrics. The efficiencies of balanced 8-harness satin fabric composites, compared to [0/90] laminates made from uni-

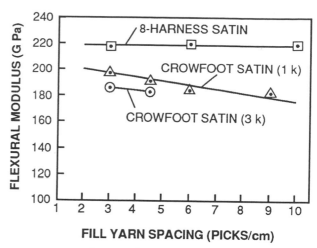

FIGURE 1.2-24. Influence of fabric construction on composite flexural modulus (data normalized to 100% warp fiber volume fraction).

directional plies, were studied by Zweben and Norman [52]. Table 1.2-1 presents data for tensile modulus and tensile and compressive strengths for composites reinforced with graphite and with aramid (Kevlar 49), and two types of hybrids having aramid/graphite fiber ratios of 50/50 and 25/75. We observe that efficiencies appear to be strongly fiber dependent, as aramid fabrics tend to

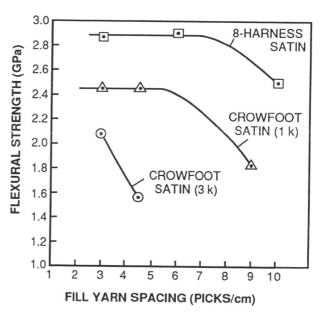

FIGURE 1.2-25. Influence of fabric construction on composite flexural strength (data normalized to 100% warp fiber volume fraction).

Table 1.2-1. Efficiency of graphite, aramid and hybrid fabric composites compared to [0/90] laminates made from unidirectional layers (data normalized to 65% fiber volume fraction).

Ratio of Aramid-to-Graphite Fiber	Tensile Modulus			Tensile Strength			Compression Strength		
	[0/90] (GPa)	Fabric (GPa)	Fabric Efficiency (%)	[0/90] (MPa)	Fabric (MPa)	Fabric Efficiency (%)	[0/90] (MPa)	Fabric (MPa)	Fabric Efficiency (%)
100/0	36.5	35.8	98	579	544	94	165	152	92
50/50	55.1	48.2	87	572	400	70	407	227	56
25/75	69.6	57.2	82	661	434	66	641	317	49
0/100	72.3	59.9	83	730	434	59	965	558	58

have mechanical properties close to those of [0/90] laminates, while graphite and hybrid efficiencies are much lower. We note that, as expected, modulus efficiencies are significantly greater than strength efficiencies for graphite and hybrid fabrics.

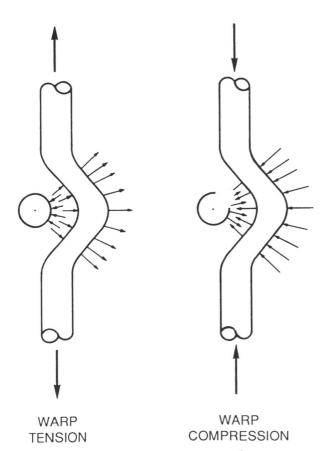

WARP
TENSION

WARP
COMPRESSION

FIGURE 1.2-26. Schematic representation of stresses on a crimped fiber loaded in tension and compression.

As noted earlier, yarn crimp does affect composite properties significantly. During weaving, tension in warp yarns is generally greater than that in the picks. Consequently, the amount of crimp in these yarns differs, and this results in differences between composite warp and fill properties, even for nominally balanced fabrics. There are other sources of warp and fill property differences, such as yarn spacing variability and skewing of yarns, which will be discussed further later.

Failure mechanisms in fabric composites have not been studied as extensively as those in unidirectional composites. However, some prominent aspects can be identified. The most important feature is crimp.

Figure 1.2-26 shows schematically the stresses acting on a yarn subjected to tensile and to compressive axial stresses. For ease of discussion, we assume the loaded yarn is a warp end. Under tensile loading, the yarn tends to straighten out. This results in compressive stresses on the fiber and in the matrix between the warp yarn and the adjacent fill yarn, which is shown in cross section. The compressive stress in the matrix acts on the fill yarn, tending to displace it laterally. This phenomenon was observed by Zweben and Wardle [53] in flexural fatigue tests on woven roving laminates. Fill yarns on the bottom surface, where warp yarns are in tension, were pushed outward, creating a checkerboard pattern that pulsated in phase with the applied load. The stress on the convex part of the crimped warp fiber is tensile, so that the fiber tends to debond from the matrix in this region. This debonding was also observed in reference [53]. We note that the actual stress state in the vicinity of fiber curvature is extremely complex. Our discussion here greatly simplifies the situation to emphasize the major features.

Next, consider a warp fiber loaded in compression. The fiber tends to buckle outward in the crossover region because of the curvature. Buckling is resisted by compressive matrix stress on the outer, convex surface and tensile stress on the inner, concave surface. The interfacial tensile stress promotes fiber debonding in the latter region. This, in turn, reduces the lateral constraint on the fiber, intensifying its tendency to buckle. This phenomenon was observed by Zweben and Wardle [53] on the top surface of woven roving flexural specimens, where warp yarns are in compression. The authors found that the addition of a layer of mat on the compressive surface of a woven roving laminate raised the compressive stress at which fiber buckling occurred, apparently by providing additional lateral restraint to the curved fibers, thereby raising their buckling stress. Further aspects of strength of fabric composites are discussed in section 1.5.

To summarize, the effect of crimp is to produce local stress perturbations which tend to promote fiber debonding, and other local failure mechanisms. In compression, crimp also promotes local fiber buckling, which results in lower laminate compressive strength compared to [0/90] composites made from unidirectional layers.

The final aspect of fabric mechanics concerns the effect of skewing (the nonorthogonality of warp and fill yarns) which arises from several sources, including nonuniform yarn tension during weaving and handling during operations such as prepregging, flaming, and scouring (processes used to remove weave finish) and lay-up.

Consider, for example, a fabric in which the fill yarns are not perpendicular to the warp yarns (Figure 1.2-27). As the figure shows, when an orthogonal balanced fabric is subjected to a temperature change, it undergoes a uniform expansion or contraction, and warp and fill yarns remain orthogonal. A square figure drawn on the fabric with its sides parallel and perpendicular to the warp direction remains a square, with longer or shorter sides than the originals. However, when a skewed fabric is subjected to a temperature change, a square with its sides originally parallel to and perpendicular to the warp distorts into a rhombus, the sides of which are also longer or shorter than the original figure. This angular distortion can, and frequently does, cause warping of parts during cool down after elevated temperature cure.

Because of the lack of symmetry, a skewed fabric also undergoes an angular distortion when it is loaded along the warp direction. This phenomenon is known as

tension-shear coupling and is discussed in Volume 2 (section 2.2).

1.2.4 Discontinuous Fiber Composites

Discontinuous fibers are the most widely used type of reinforcement. As a rule, fibers are intended to be randomly distributed throughout the material. In practice, this may be far from the case because many aspects of fabrication processes, especially flow, tend to align fibers. When fiber length is smaller than part thickness, fibers can be oriented in three dimensions. When it is greater, fibers tend to lie primarily in the plane of the part. We refer to these cases as three-dimensional and two-dimensional arrangements, respectively. Figures 1.2-28 and 1.2-29 illustrate these two kinds of materials.

When fibers are randomly distributed in three dimensions, there is no preferred direction as there is for unidirectional composites, and the former are statistically isotropic. That is, on the average, the properties are the same in every direction. Two-dimensional arrangements with randomly oriented fibers are isotropic in the plane of the material; however, the properties in the thickness direction are different from those in the plane. A material with these properties is sometimes referred to as "quasi-isotropic."

In the discussion of factors affecting the properties of unidirectional composites we noted that fiber volume fraction straightness, orientation, and interfacial bond strength characteristics are important. These parameters are of particular significance for discontinuous fiber composites. There are some additional factors that

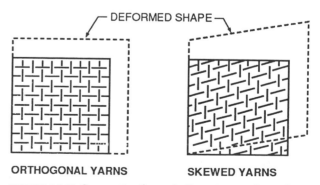

FIGURE 1.2-27. Composite thermal distortion resulting from skewed fabric.

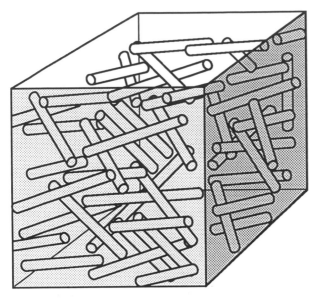

FIGURE 1.2-28.

have major influences for discontinuous fiber composites, mainly fiber length and diameter. Fibers are often distributed throughout a composite in the form of fiber bundles, i.e., chopped ends, yarns or rovings. For these cases, the effective diameter of the fiber bundle is also important. The influence of the parameters mentioned above will be considered in this subsection.

We have emphasized throughout this section that variability, which is an important characteristic of composites, is particularly true for discontinuous fiber composites because major parameters like fiber orientation and local fiber volume fraction are difficult to control in most processes. In addition to the inherent property variability of composites with randomly oriented fibers, fabrication processes often result in regions of preferred fiber orientation and low fiber content. This process-induced material variability must be considered to assure reliable performance. Of course, every effort should be made to reduce this problem

by careful attention to the details of the fabrication method.

In considering fiber volume fraction, it is important to keep in mind the distinction between the average fiber volume fraction of the material and the local fiber volume fraction in a particular region. For example, consider a composite reinforced with chopped rovings that are not broken up to disperse individual fibers. The local fiber volume fraction in the fiber bundle can be relatively high, say 50 to 60%, while the average volume fraction of the material will be much lower. The range of average fiber volume fractions for discontinuous fiber composites is greater than those of materials using other forms of reinforcement. For example, the fiber volume fraction of some injection molded materials is less than 10%, and the value for some sheet molding compounds is as great as 50%. Generally, volume fractions tend to be lower than for unidirectional and fabric composites, and this contributes to the lower mechanical property values of discontinuous fiber composites.

In the discussion of unidirectional composites, we emphasized that these materials are strongly anisotropic, and properties vary greatly with direction. Modulus and strength properties in the fiber direction are much greater than those in the transverse direction. Strength and stiffness tend to drop off quickly as the angle between the fiber and loading directions increases. In a composite reinforced with randomly distributed fibers, only a small percentage contributes significantly to strength and stiffness when load is applied in any one direction. Consequently, the tensile and compressive strengths and extensional modulus of random fiber composites are much lower than the corresponding axial properties of unidirectional composites having the same fiber volume fractions. However, random fiber composite mechanical properties are usually substantially greater than the transverse properties of unidirectional composites. Later in this subsection, we present some simple expressions relating the modulus and strength of random fiber composites to those of their constituents. We noted previously that fiber curvature reduces reinforcement efficiency—a significant effect in discontinuous fiber composites made by processes in which there is a significant amount of curvature induced by flow. As in the cases of regions with low fiber content or significant fiber orientation, regions containing a significant amount of fiber curvature are likely to be possible sources of part failure because of their lower strength properties.

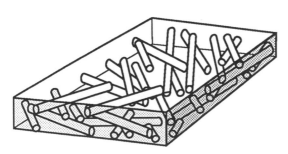

FIGURE 1.2-29.

The parameters we have discussed above, fiber volume fraction, orientation and curvature, are important for all types of composites. We now turn our attention to the influence of fiber length which is unique to discontinuous fiber composites.

Discontinuous Fiber Aligned with Load

Consider the case of a fiber of length ℓ and diameter d imbedded in a composite material subjected to tensile loading parallel to the fiber direction, as shown in Figure 1.2-30. As discussed above, fibers are often not individually dispersed throughout the material but are, instead, incorporated into the composite in bundles such as chopped ends and rovings. In this case, we assume that the bundle itself behaves as if it were a fiber, and we use the effective diameter of the bundle for d. As we shall see, fiber aspect ratio ℓ/d is a significant parameter for discontinuous fiber composites. The effective diameter of fiber bundles is generally at least an order of magnitude greater than that of a single fiber. For convenience, in this discussion we will use the term *fiber* with the understanding that it refers to both individual fibers and fiber bundles.

We assume that the average strain in the composite parallel to the load direction is e. The local strains in the material differ substantially from this value because of the complex state of internal stress. There are only two mechanisms by which the fiber can be loaded in axial tension: by tensile stress on the ends of the fiber or by axial shear stress on the cylindrical surface of the fiber. Therefore, the ability of the fiber to resist axial tensile load depends on the interfacial bond strength between the fiber and matrix. If the fiber were coated with a lubricant so that stress transfer from the matrix to the fiber were impossible, it would not contribute to the strength or stiffness of the material. In fact, the effect would be that of a cylindrical cavity of length ℓ and diameter d in the material. As a result, the stiffness and strength of the composite would be lower than that of the resin alone.

Several authors have studied the case of stress transfer between fibers and matrix in composites reinforced with discontinuous fibers using models with varying degrees of complexity [54–58]. The problem is related to that of stresses near broken continuous fibers discussed in references [23–25]. We will not attempt to discuss these analyses here. Instead, we present a simplified synthesis of the problem, emphasizing the major parameters and phenomena involved.

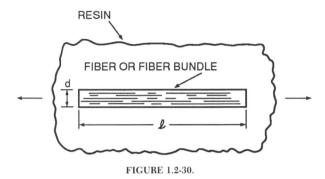

FIGURE 1.2-30.

Following Cox [54] we represent the discontinuous fiber in a composite by a simple model consisting of a circular cylindrical fiber of radius $r = d/2$ and length ℓ imbedded in a circular cylinder of matrix whose outer radius is r_m, as shown in Figure 1.2-31. We suppose initially that the matrix and fiber are both elastic and perfectly bonded together. Load transfer between the matrix and fiber resulting from tensile stress on the fiber ends is neglected. Figure 1.2-32 shows the variation of the fiber tensile stress, σ, and the shear stress at the fiber-matrix interface, τ, with the distance from the end of the fiber, x. σ is zero at the fiber ends and increases to a maximum, σ_{max}, at the center of the fiber. For very long fibers, σ_{max} approaches $E_f e$, the stress that would exist in a continuous fiber subjected to the composite strain, e.

There are two major results of interest. First, we note that the fiber is not fully stressed over its entire length, as a continuous fiber would be. This means that discontinuous fibers are less efficient reinforcements. Second, there is a shear stress concentration at the ends of the fibers. As we shall see, these shear stresses can be large, and they have an important effect on strength.

Let us consider the question of reinforcing efficiency in more detail. Our concern at this point is with modulus. We will consider strength later. The contribution of a fiber to composite modulus depends on its average tensile stress, $\bar{\sigma}$. For a continuous fiber, stress is constant, so that $\bar{\sigma} = E_f e$. The average stress in a discontinuous fiber is less than this value. We denote the ratio

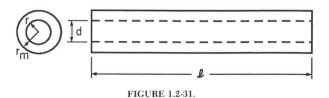

FIGURE 1.2-31.

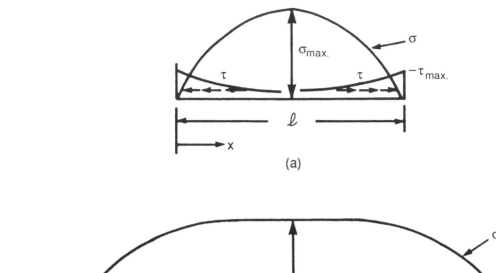

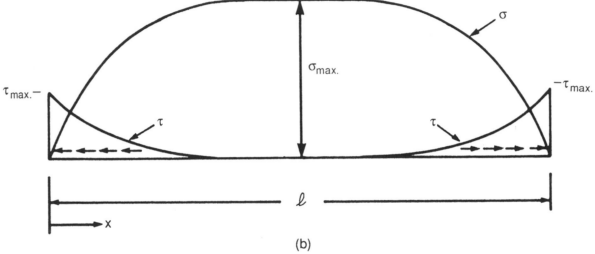

FIGURE 1.2-32.

of average stress in a discontinuous fiber to that of a continuous fiber for the elastic case by ϕ_e, which we call the modulus reinforcement efficiency. If we define the radius of the matrix cylinder shown in Figure 1.2-31 so that $v_f = \pi r^2 / \pi r_m^2$, then Cox's expression for modulus reinforcement efficiency can be put in the following form:

$$\phi_e = 1 - \frac{\tanh p}{p} \qquad (1.2\text{-}25)$$

where

$$p = 2\left(\frac{\ell}{d}\right)\left(\frac{G_m}{E_f}\right)^{1/2}\left(\frac{-1}{\ln v_f}\right)^{1/2} \qquad (1.2\text{-}26)$$

and ln is the natural logarithm.

We see from Equation (1.2-25) that as p goes to zero, so does fiber modulus reinforcing efficiency. Conversely, as p approaches infinity, ϕ_e goes to unity, and the discontinuous fiber is almost as efficient from a modulus standpoint as a continuous one. Equation (1.2-26) shows us on which material parameters p depends. We see that p varies directly with the fiber aspect ratio, ℓ/d, which seems reasonable because we would expect a discontinuous fiber to become more efficient as its length increases. p also depends on $(G_m/E_f)^{1/2}$, so that for a given fiber length, diameter, and volume fraction, reinforcement *efficiency* decreases as fiber modulus increases. Note that this does not imply that the absolute value of composite modulus decreases with increasing fiber stiffness. It simply states that the efficiency of discontinuous fibers decreases with increasing stiffness.

We can use the results presented above to obtain an approximate expression for the longitudinal modulus of a composite reinforced with parallel discontinuous fibers:

$$E_c = \phi_e v_f E_f + (1 - v_f) E_m \qquad (1.2\text{-}27)$$

This formula applies when the stresses in the fiber and matrix are in the elastic range.

We would anticipate that the contribution of aligned discontinuous fibers to the transverse and shear moduli should be similar to that of continuous fibers unless the former have a very small aspect ratio.

As we mentioned earlier, our primary interest at this point lies in the consideration of modulus reinforcing efficiency and the effects of shear stress at fiber ends. We now consider the latter effect.

As Figure 1.2-32 shows, interfacial shear stress is concentrated at the ends of discontinuous fibers, as it is for broken continuous fibers (see subsection 1.2.2). Kelly [22] used Cox's analysis to obtain an expression for the ratio of maximum shear stress-to-fiber tensile stress, τ_{max}/σ_{max}, which we present in a modified form to emphasize the physical parameters involved:

$$\frac{\tau_{max}}{\sigma_{max}} = \left(\frac{G_m}{E_f} \right)^{1/2} \left(\frac{-1}{\ln v_f} \right)^{1/2} \coth \frac{p}{2} \qquad (1.2\text{-}28)$$

For a very long fiber, $\coth p/2$ approaches unity, therefore:

$$\frac{\tau_{max}}{\sigma_{max}} = \left(\frac{G_m}{E_f} \right)^{1/2} \left(\frac{-1}{\ln v_f} \right)^{1/2} \qquad (1.2\text{-}29)$$

For a composite consisting of 50% E glass fibers by volume in a typical epoxy or polyester resin, the ratio of maximum shear stress-to-tensile stress predicted by Equation (1.2-29) is 0.19. This result means that the maximum shear stress at the ends of a discontinuous fiber is about 20% of the maximum fiber tensile stress. This stress is quite high considering that matrix shear strengths and fiber-matrix interfacial shear strengths are very low compared to fiber tensile strengths. Experiments have shown that actual shear stresses are much higher than predicted by this simple analysis [58]. Localized matrix failure or fiber debonding undoubtedly occurs in most discontinuous fiber composites as they are loaded to failure.

Equation (1.2-29) shows that for very long fibers,

τ_{max}/σ_{max} decreases as E_f increases. Referring to Equation (1.2-28), this is generally true for shorter fibers because the variation of $\coth p/2$ with E_f is weaker than $E_f^{1/2}$. The physical interpretation of this dependence is that as the fiber gets stiffer, the shear stress that loads the fiber at its ends is spread over a longer distance, and its maximum intensity is reduced. Because of the reduced shear stress intensity, it takes a greater length of fiber to reach a given stress level for a fiber of high stiffness than for one of low stiffness. Thus, modulus reinforcing efficiency decreases with increasing fiber modulus.

So far we have considered the case of an elastic fiber perfectly bonded to an elastic matrix. We obtained expressions for modulus reinforcement efficiency and maximum shear stress, which occurs at the fiber ends. We found that the maximum shear stress predicted by the model is high and noted that experimental data show that actual stresses are much higher. Consequently, as experimental observations indicate, we expect that some sort of localized matrix failure, interfacial failure, or both will occur at fiber ends before composite failure takes place. When localized failure takes place, the elastic model used so far is no longer valid; consequently, the expression for fiber modulus efficiency generally will not be valid for fiber strength efficiency. We now consider the effects of matrix and bond failure at fiber ends.

The actual state of stress at the ends of a discontinuous fiber is more complex than that represented by the simplified model just discussed. In fact, the sharp edges give rise to stress singularities. The situation is further complicated when localized failure occurs. There are several possible failure modes that have been identified, such as matrix cracking perpendicular to the fiber, fiber-matrix interfacial failure and plastic yielding of the matrix [22]. When the interfacial strength is low compared to matrix strength properties, as is frequently the case, debonding will occur before matrix failure. Figure 1.2-33 shows some idealized interfacial shear stress distributions that can occur at the ends of a discontinuous fiber when the applied load is increased beyond the level that caused the maximum interfacial shear stress τ_{max} to reach the interfacial shear strength, τ_u. Failure is initiated at the ends of the fiber and the failure zone progresses inward toward the center of the fiber as the applied load is increased. When the interfacial bond is broken, shear stress transfer will result primarily from friction. For simplicity, we have pre-

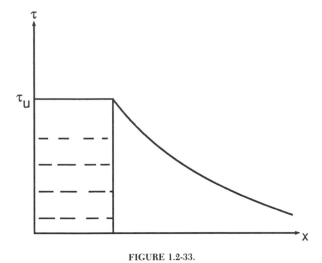

FIGURE 1.2-33.

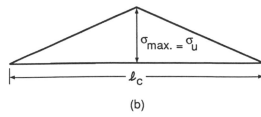

(a)

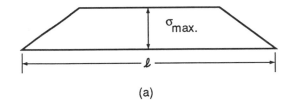

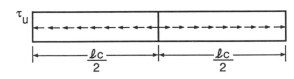

(b)

FIGURE 1.2-34.

sented the stress in the failed region as being constant and equal to some fraction of the failure stress τ_u. If the stress drops to zero, the debonded region can propagate unstably for a small load increment [23,24]. At the other end of the spectrum, the shear stress equals failure stress τ_u, as indicated by the solid line. This equivalence is the same as the case of a perfectly bonded discontinuous fiber in an elastic, perfectly plastic material. For a matrix with work hardening, the stress in the failed region can exceed τ_u.

Although the case of constant interfacial shear stress is highly idealized, it provides valuable insights into the influence of nonelastic effects initiating at fiber ends on the behavior of discontinuous fiber composites. This influence is important, as stresses at fiber ends are high, and interfacial shear strengths frequently are lower than matrix strengths. The particular approach presented is that proposed by Kelly [22].

Consider a discontinuous fiber that has been loaded past the point of interfacial or matrix failure. Assume that the interfacial shear stress in the failed region is constant and equal to τ_u. We completely neglect the contribution of the elastic portion of the shear stress on the fiber. Figure 1.2-34 shows the idealized state of shear stress and resulting fiber tensile stress assumed in this model. The maximum tensile stress in the fiber, σ_{max}, can be calculated from the total shear force acting on the fiber ends:

$$\frac{(\pi d^2)}{4}\sigma_{max} = (\pi d)\ell_T\tau_u \qquad (1.2\text{-}30)$$

where ℓ_T, the transfer length, is the length of fiber over which the constant shear stress acts at the end of a fiber. As discussed above, Equation (1.2-25) for modulus reinforcement efficiency, ϕ_e, is based on an elastic analysis that is not valid when matrix or interfacial failure occurs near the fiber ends. Therefore, we return to the basic definition of reinforcement efficiency as the ratio of mean stress in a discontinuous fiber to that in a continuous fiber. This definition gives us an expression for ϕ_p, the reinforcement efficiency after localized failure has occurred:

$$\phi_p = 1 - \frac{\ell_T}{\ell} \qquad (1.2\text{-}31)$$

Solving Equation (1.2-30) for ℓ_T, we find:

$$\phi_p = 1 - \frac{1}{4}\left(\frac{d}{\ell}\right)\frac{\sigma_{max}}{\tau_u} \qquad (1.2\text{-}32)$$

We are interested in the limiting case where the fiber tensile stress reaches its ultimate strength value, σ_u. We designate the value of ϕ_p at this point by ϕ_u, which we refer to as the strength reinforcing efficiency. It is given by:

$$\phi_u = 1 - \frac{1}{4}\left(\frac{d}{\ell}\right)\frac{\sigma_u}{\tau_u} \qquad (1.2\text{-}33)$$

Observe that the fiber aspect ratio ℓ/d appears in expressions for ϕ_e, ϕ_p and ϕ_u. Obviously, this is a very important parameter for discontinuous fiber composites. According to this model, reinforcement efficiency after localized failure at fiber ends depends strongly on the interfacial shear stress, τ_u. Anything that reduces interfacial shear stress transfer, such as the possible influence of moisture, reduces fiber reinforcing efficiency.

According to Equation (1.2-33), strength reinforcing efficiency decreases with increasing fiber strength because the failed region of length ℓ_T at the end of the fibers grows as σ_{max} increases [Equation (1.2-30)]. The progressive reduction in fiber efficiency after localized failure at the fiber ends means that the contribution of the fibers to composite modulus steadily decreases; this decrease is observed experimentally. The tangent modulus of discontinuous fiber composites usually decreases above some proportional limit, as shown schematically in Figure 1.2-35. However, there can be other contributions to the observed nonlinearity, such as the nonlinear stress-strain behavior of the matrix itself. In composites with randomly oriented fibers, transverse failure modes associated with fibers or fiber bundles that are not aligned with the applied load can also result in a departure from linear behavior.

Randomly Oriented Discontinuous Fibers

We have limited our discussion to the case where all of the fibers are parallel and aligned with the applied load. We now consider the more general case where fibers are randomly oriented, both in a plane, and in three dimensions. We consider modulus first.

A number of papers have analyzed the problem of the modulus of composites with randomly oriented fibers [54,59–63]. Perhaps the simplest result to use is that of Cox [54] which we can combine with his results for the efficiency of discontinuous fibers discussed earlier. This combination yields the following expression for the modulus of a composite reinforced with a planar

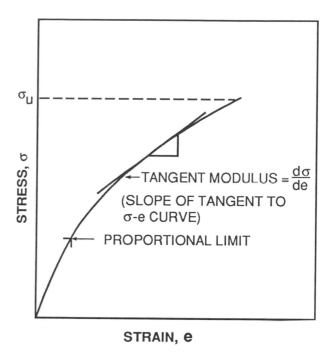

FIGURE 1.2-35.

(two-dimensional) arrangement of randomly oriented discontinuous fibers:

$$E_c = \frac{1}{3}\phi_e V_f E_f + (1 - V_f)E_m \qquad (1.2\text{-}34)$$

Comparing this expression with Equation (1.2-27) we see that they are the same, except for the factor of 1/3 in the first term on the right. This factor represents the influence of the random fiber orientation; fiber contribution to modulus is only one-third as great as when they are all aligned. However, the modulus of a random-fiber composite is the same for all directions in the plane, whereas that of an aligned-fiber composite is lower in the transverse direction.

For the case where fibers are randomly oriented in three dimensions, the modulus is approximated by:

$$E_c = \frac{1}{6}\phi_e V_f E_f + (1 - V_f)E_m \qquad (1.2\text{-}35)$$

For this class of materials, the predicted contribution of the fibers is only one-sixth that of the aligned case. This fact, along with low fiber volume fraction and perhaps the short lengths used, explains why the modulus of materials with three-dimensional reinforcement, such as injection molded plastics, is low compared to two-dimensional laminates with aligned continuous fibers.

In subsection 1.2.2, we saw that unidirectional composite failure modes are extremely complex, even under simple loading conditions such as axial tension. The failure process for composites with randomly oriented discontinuous fibers is more complicated. Because fibers are oriented in all directions, both transverse and axial failure modes can contribute to composite failure [63] by transferring load at fiber ends; therefore, matrix and interfacial strengths are of critical importance.

1.2.5 References

1. WHITNEY, J. M. and R. J. Nuismer. "Stress Fracture Criteria for Laminated Composites Containing Stress Concentrations," *J. Composite Materials*, 8:253 (1974).

2. GERE, J. M. and S. P. Timoshenko. *Mechanics of Materials, 2nd ed.* PWS Publishers, Boston (1984).

3. JONES, R. M. *Mechanics of Composite Materials*. McGraw-Hill Book Co., Boston (1975).

4. HILL, R. "Elastic Properties of Reinforced Solids: Some Theoretical Principles," *Journal of the Mechanics and Physics of Solids*, 11:357 (1963).

5. HILL, R. "Theory of Mechanical Properties of Fibre-Strengthened Materials. I Elastic Behavior," *Journal of the Mechanics and Physics of Solids*, 12:199 (1964).

6. HILL, R. "A Self-Consistent Mechanics of Composite Materials," *Journal of the Mechanics and Physics of Solids*, 13:213 (1965).

7. HASHIN, Z. "On Elastic Behavior of Fibre Reinforced Materials of Arbitrary Transverse Phase Geometry," *Journal of the Mechanics and Physics of Solids*, 13:119 (1965).

8. HASHIN, Z. "Theory of Composite Materials," *Mechanics of Composite Materials*, F. W. Wendt, H. Liebowitz, N. Perrone, eds., Pergamon Press, Oxford, p. 201 (1970).

9. HASHIN, Z. and B. W. Rosen. "The Elastic Moduli of Fiber Reinforced Materials," *Journal of Applied Mechanics*, 31:223 (1964).

10. PAUL, B. "Prediction of Elastic Constants of Multiphase Materials," *AIME Transactions*, 219:36 (1960).

11. CHAMIS, C. C. and G. P. Sendeckyj. "Critique on Theories Predicting Thermoelastic Properties of Fibrous Composites," *Journal of Composite Materials*, 2:332 (1968).

12. ADAMS, D. F. and S. W. Tsai. "The Influence of Random Filament Packing on the Elastic Properties of Composite Materials," Memorandum RM-5608-PR, The Rand Corporation (December 1968).

13. ADAMS, D. F. and D. R. Doner. "Transverse Normal Loading of a Unidirectional Composite," *Journal of Composite Materials*, 1:152 (1967).

14. FOYE, R. L. "An Evaluation of Various Engineering Estimates of the Transverse Properties of Unidirectional Composites," *Proceedings of the Tenth National SAMPE Symposium* (November 1966).

15. ADAMS, D. F., D. R. Doner and R. L. Thomas. "Mechanical Behavior of Fiber-Reinforced Composite Materials," AFML-TR-67-96, U.S. Air Force Materials Laboratory (May 1967).

16. ASTM D2344-84, "Standard Test Method for Apparent Interlaminar Shear Strength of Parallel Fiber Composites by Short-Beam Method."

17. ZWEBEN, C. "Tensile Failure Analysis of Fibrous Composites," *AIAA Journal*, 6(12):2325 (1968).

18. ZWEBEN, C. "A Bounding Approach to the Strength of Composite Materials," *Engineering Fracture Mechanics*, 4:1 (1972).

19. ROSEN, B. W. "Tensile Failure of Fibrous Composites," *AIAA Journal*, 2(11):1982 (1964).

20. LIPTAI, R. G. "Acoustic Emission from Composite Materials," *Composite Materials: Testing and Design (Second Conference)*, ASTM STP 497, American Society for Testing and Materials, p. 285 (1972).

21. FRIEDMAN, E. "A Tensile Failure Mechanism for Whisker Reinforced Plastics," *Proceedings of the Twenty-Second Annual Conference of the SPI Reinforced Plastics/Composites Institute*, Washington, D.C. (February 1967).

22. KELLY, A. *Strong Solids, Second Edition*. Clarendon Press, Oxford (1973).

23. HEDGEPETH, J. M. and P. Van Dyke. "Local Stress Concentrations in Imperfect Filamentary Composite Materials," *Journal of Composite Materials*, 1:294 (1967).

24. ZWEBEN, C. "An Approximate Method of Analysis for Notched Unidirectional Composites," *Engineering Fracture Mechanics*, 6:1 (1974).

25. HEDGEPETH, J. M. "Stress Concentrations in Filamentary Structures," NASA TN D-882, National Aeronautics and Space Administration (May 1961).

26. VAN DYKE, P. and J. M. Hedgepeth. "Stress Concentrations from Single-Filament Failures in Composite Materials," *Textile Research*, 39(7):618–626 (1969).

27. ZWEBEN, C. and B. W. Rosen. "A Statistical Theory of Material Strength with Application to Composite Materials," *Journal of the Mechanics and Physics of Solids*, 18:189 (1970).

28. SCOP, P. M. and A. S. Argon. "Statistical Theory of Strength of Laminated Composites," *Journal of Composite Materials*, 1:92 (1967).

29. PHOENIX, S. L. "Probabilistic Concepts in Modeling the Tensile Strength Behavior of Fiber Bundles and Unidirectional Fiber Matrix Composites," *Composite Materials: Testing and Design*, ASTM STP 546, American Society for Testing and Materials, Philadelphia, p. 130 (1974).

30. HARLOW, D. G. and S. L. Phoenix. "The Chain-of-Bundles Probability Model for the Strength of Fibrous Materials, I: Analysis and Conjectures," *Journal of Composite Materials*, 12:195 (1978).

31. MULLIN, J. V. and V. F. Mazzio. "The Effects of Matrix and Interface Modification on Local Fractures of Carbon Fibres

in Epoxy," *Journal of the Mechanics and Physics of Solids*, 20:391 (1972).

32. WEIBULL, W. "A Statistical Theory of the Strength of Materials," *Ing. Vetenskaps, Akad. Handl.*, NR 151 (1939).

33. CROWTHER, M. F. and M. S. Starkey. "Use of Weibull Statistics to Quantify Specimen Size Effects in Fatigue of GRP," *Composites Science and Technology*, 31:87 (1988).

34. BULLOCK, R. E. "Strength Ratios of Composite Materials in Flexure and Tension," *Journal of Composite Materials*, 8:200–206 (1974).

35. ZWEBEN, C. "The Flexural Strength of Aramid Fiber Composites," *Journal of Composite Materials*, 12:422 (1978).

36. RIEDINGER, L. A., M. H. Kural and G. W. Reed, Jr. "Evaluation of the Potential Structural Performance of Composites," *Mechanics of Composite Materials*, F. W. Wendt, H. Liebowitz and N. Perrone, eds., Pergamon Press, Oxford (1970).

37. GREENWOOD, J. H. and P. G. Rose. "Compressive Behavior of Kevlar 49 Fibres and Composites," *Journal of Materials Science*, 9:1809 (1974).

38. KULKARNI, S. V., J. S. Rice and B. W. Rosen. "An Investigation of the Compressive Strength of Kevlar 49/Epoxy Composites," *Composites*, 6:217 (1975).

39. DETERESA, S. J., R. S. Porter and R. J. Farris. "Experimental Verification of a Microbuckling Model for the Axial Compressive Failure of High Performance Polymer Fibers," *J. Materials Science*, 23:1851 (1988).

40. SCHNERCH, H. "Prediction of Compressive Strength in Uniaxial Boron Fiber-Metal Matrix Composite Materials," *AIAA Journal*, 4 (1966).

41. ROSEN, B. W. "Mechanics of Composite Strengthening," *Fiber Composite Materials*, American Society for Metals, Metals Park, Ohio (1965).

42. EVANS, A. G. and W. F. Adler. "Kinking as a Mode of Failure in Carbon Fiber Composites," *Acta Metallurgica*, 26:725 (1978).

43. DAVIS, J. G. "Compressive Strength of Lamina Reinforced and Fiber Reinforced Composite Materials," Ph.D. Thesis, Virginia Polytechnic Institute and State University (1973).

44. Brelant, S. and I. Petker. "Fabrication and Environmental Interaction Effects of Filament-Wound Composites," *Mechanics of Composite Materials*, F. W. Wendt, H. Liebowitz and N. Perrone, eds., Pergamon Press, Oxford (1970).

45. YOSHIDA, H., T. Ogasa and R. Hayashi. "Statistical Approach to the Relationship Between ILSS and Void Content of CFRP," *Composites Science and Technology*, 25:331 (1986).

46. ADAMS, D. F. and D. R. Doner. "Longitudinal Shear Loading of a Unidirectional Composite," *Journal of Composite Materials*, 1:4 (1967).

47. WOODBERRY, R. F. H. and D. E. Borgmeier. "Application of Advanced Fibers to Chambers for Solid Propellant Rocket Motors," Chemical Propulsion Information Agency Meeting, Silver Springs, Maryland (November 1973).

48. KO, F. K. "Three-Dimensional Fabrics for Composites—An Introduction to the Magnaweave Structure," in *Progress in Science and Engineering of Composites*, Hayashi et al., eds., ICCM-IV, Tokyo, p. 1069 (1982).

49. MA, C. L., J. M. Yang and T. W. Chou. "Elastic Stiffness of Three-Dimensional Braided Textile Structural Composites," in *Composite Materials: Testing and Design*, ASTM STP 893, p. 404 (1986).

50. SHIBATA, N., A. Nishimura and T. Norita. "Graphite Fiber's Fabric Design and Composite Properties," *SAMPE Quarterly*, p. 25 (July 1976).

51. ZWEBEN, C., W. S. Smith and M. W. Wardle. "Test Methods for Fiber Tensile Strength, Composite Flexural Modulus and Properties of Fabric-Reinforced Laminates," *Composite Materials: Testing and Design—Fifth Conference*, ASTM STP 674, American Society for Testing and Materials, p. 228 (1978).

52. ZWEBEN, C. and J. C. Norman. "Kevlar 49/Thornel 300 Hybrid Fabric Composites for Aerospace Applications," *SAMPE Quarterly*, p. 1 (July 1976).

53. ZWEBEN, C. and M. W. Wardle. "Flexural Fatigue of Marine Laminates Reinforced with Woven Roving of E-Glass and of Kevlar 49 Aramid," *Proceedings of the Thirty-Fourth Annual Meeting of the SPI Reinforced Plastics/Composites Institute, New Orleans* (February 1979).

54. COX, H. L. "The Elasticity and Strength of Paper and Other Fibrous Materials," *British Journal of Applied Physics*, 3:72 (1952).

55. OUTWATER, J. P., JR. "The Mechanics of Plastics Reinforcement in Tension," *Modern Plastics*, 33:156 (1956).

56. DOW, N. F. "Study of Stresses Near a Discontinuity in a Filament-Reinforced Composite Material," Report No. TIS R63SD61, General Electric Company (August 1963).

57. SMITH, G. E. and A. J. M. Spencer. "Interfacial Tractions in a Fibre-Reinforced Elastic Composite Material," *Journal of the Mechanics and Physics of Solids*, 18:81 (1970).

58. TYSON, W. R. and G. J. Davies. "A Photoelastic Study of the Shear Stresses Associated with the Transfer of Stress During Fibre Reinforcement," *British Journal of Applied Physics*, 16:199 (1965).

59. TSAI, S. W. and N. J. Pagano. "Invariant Properties of Composite Materials," *Composite Materials Workshop*, S. W. Tsai, J. C. Halpin and N. J. Pagano, eds., Technomic Publishing Co., Inc., Lancaster, PA (1968).

60. CHRISTENSEN, R. M. and F. M. Waals. "Effective Stiffness of Randomly Oriented Fiber Composites," *Journal of Composite Materials*, 6:518 (1972).

61. WU, C.-T. D. and R. L. McCullough. "Constitutive Relationships for Heterogeneous Materials," *Developments in Composite Materials—1*, G. S. Hollister, ed., Applied Science Publishers, Ltd., London (1977).

62. HALPIN, J. C. and N. J. Pagano. "The Laminate Approximation for Randomly Oriented Fibrous Composites," *J. Composite Materials*, 3:720 (1969).

63. HALPIN, J. C. and J. L. Kardos. "Strength of Discontinuous Reinforced Composites: I. Fiber Reinforced Composites," *Polymer Engineering and Science*, 18:496 (1978).

Static Strength and Elastic Properties

1.3 STATIC STRENGTH AND ELASTIC PROPERTIES

C. ZWEBEN

1.3.1 Introduction

In this section we consider elastic and static strength properties of resin-matrix composites reinforced with continuous unidirectional fibers, fabrics, discontinuous fibers and particulates. The objectives of this section are to provide the designer with:

- representative property data for preliminary design and material trade-off studies
- methods for dealing with the critical problem of property variability
- procedures for assessing the validity of composite property data

As discussed in section 1.2, composite properties are strongly dependent on fabrication processes and the particular material systems used. Therefore, efficient and reliable designs should be based on data obtained from test specimens made by methods that closely reproduce those that will be used for the final components. In deciding on design allowables, the engineer should also be aware of the fact that there is considerable variation in test methods throughout the industry. Consequently, the procedures by which mechanical property data are determined should be evaluated. The subject of test methods is treated in detail in Volume 6.

Although elastic properties and static strength characteristics are important for design, they are only part of the story. Fatigue, creep, relaxation, failure under sustained loading (frequently called static fatigue or creep rupture) and environmental effects are also important design considerations. These topics will be treated elsewhere in Volume 1.

In this section, we examine two important aspects of composite mechanical properties as they relate to design: the concept of effective mechanical properties and property variability. We then present an overview of mechanical property data for key composite ma-terials and reinforcing forms, along with comparative property data for metals. Some of the definitions of material properties from section 1.2 are repeated for convenience.

1.3.2 An Overview of Material Properties

In this subsection, we will first define which material properties are considered "basic" and how effective properties may be used for a heterogeneous composite. We then examine the nature and sources of variability for composites reinforced with aligned continuous fibers, fabrics, and random, discontinuous fibers. A method for dealing with property variability through the use of design allowables is discussed. Finally, the possibility of a dependence of strength on volume (size effect) will be considered, and a simple way to estimate reduced allowables is presented.

Effective Properties—A Definition

We have emphasized throughout this section that composites are strongly heterogeneous. Because fibers, particulates and resins have very different elastic properties, the internal stress distribution in a composite is very complex and far from uniform, even under simple loading conditions like pure tension. However, in designing with composites we normally neglect the variation of stress and displacements on the microscopic level and consider these materials to be macroscopically homogeneous and anisotropic. We then deal with composites on the basis of effective elastic and strength properties which are either measured directly or estimated from constituent properties using various analytical or empirical formulas [1,2].

It is important to recognize the assumptions inherent in this approach and when they are valid. There are certain basic characteristic dimensions of heterogeneity, such as fiber diameter and spacing, and particulate dimensions and spacing. When discontinuous fibers are used, transfer length and fiber bundle diameter are also important characteristic dimensions. The significance of these characteristic lengths is that the stress distribution and displacement field vary greatly over volume elements having these dimensions. Consequently, we are not strictly justified in treating a structural element as a homogeneous material unless its dimensions are large with respect to the characteristic dimensions of material heterogeneity.

In practice, this dimension restriction may be ignored in a number of cases. For example, a layer of boron prepreg normally contains only one fiber through the thickness. Therefore, according to the requirement that material dimensions be large with respect to heterogeneity characteristic lengths, it is not justified to treat a single layer as if it were a homogeneous material. However, this approximation is routinely made and appears to give reasonable results. Most other fibers of interest, such as glass, graphite, and aramid, are much smaller and are produced in bundles. A layer typically contains many fibers through the thickness; therefore, it is reasonable to treat layers of composites made from them as effectively homogeneous materials. It is important to recognize, however, that a laminate is a continuum and it is impossible to say exactly where one layer ends and the next begins.

As discussed in section 1.2, fiber length has an important influence on mechanical properties. However, the question of whether it is a characteristic dimension in relation to use of effective properties is not clear cut. When discontinuous fibers are very long, the only effect of the discontinuity occurs in the vicinity of the fiber ends. The stress and displacement fields are perturbed in a region whose characteristic dimension is the transfer length. Therefore, for the case of very long fibers, fiber length does not appear to be a characteristic dimension.

When discontinuous fibers are not significantly longer than the transfer length, the states of stress and displacement vary continuously along the fiber. Consequently, the material must contain many such elements before we can truly speak of an "average" response of the material. This concept is analogous to the situation in metals where specimen dimensions must be much larger than grain size before the material can be treated as homogeneous; therefore, fiber length is a characteristic dimension when it is not very large with respect to transfer length.

In summary, strictly speaking, we are only justified in treating composites as homogeneous materials and using effective mechanical properties when the dimensions of the structural element under consideration are large with respect to the characteristic dimensions of the material heterogeneity. This rule is, however, sometimes violated without serious consequences. At present, we do not have precise guidelines to define when this violation is permissible. Therefore, the designer should use some discretion in borderline cases.

Definition of Basic Properties

By now, the reader must be well aware that fibrous composites generally are anisotropic materials, which means that their strength and elastic properties vary with direction. We discussed in detail in section 1.2 the several types of anisotropy that can exist in these materials. The number of elastic constants and strength parameters required to characterize a composite depends on the type of anisotropy.

Most of the applications for which composites are being considered are made up of relatively thin beam, plate and shell elements. Some examples are automobile hoods, trunk lids, body panels, drive shafts and springs; aircraft fairings and control surfaces; and chemical storage tanks and pressure vessels. For these structures, the in-plane extensional stresses σ_1 and σ_2, which arise from stretching and bending, are usually much greater than the extensional stress through the thickness, σ_3; therefore, the latter stress is usually neglected. (Figure 1.3-1 shows the coordinate geometry.) Further, it is usually assumed that any line perpendicular to the beam or plate midplane before deformation remains perpendicular and undeformed when load is applied. (This assumption is usually called the Kirchoff hypothesis.) As a result, the extensional strain normal to the midplane, ϵ_3, and the shear strains through the thickness, γ_{13} and γ_{23}, are zero. Based on these assumptions, we need at most four elastic constants to determine the deformation and stresses in the structure. These constants are usually taken to be:

$E_1 = E_L =$ Axial (Longitudinal) Extensional Modulus

$E_2 = E_T$ = Transverse Extensional Modulus
$G_{12} = G_{LT}$ = In-Plane Shear Modulus
$\nu_{12} = \nu_{LT}$ = Axial In-Plane Poisson's Ratio

As discussed in section 1.2, Poisson's ratio, ν_{12}, is defined as the ratio of the compressive strain in the transverse direction to the axial extensional strain when the material is loaded in tension in the axial direction. That is,

$$\nu_{12} = -\epsilon_2/\epsilon_1 \quad \text{Loading } \sigma_1 = \sigma, \ \sigma_2 = \tau_{12} = 0 \tag{1.3-1}$$

This is the most common definition of ν_{12}, however, the opposite is sometimes found in the literature. Poisson's ratio ν_{21} is related to ν_{12} by the equation:

$$\nu_{12}/\nu_{21} = E_1/E_2 \tag{1.3-2}$$

Usually, the 1-direction, or longitudinal axis, is assumed to be parallel to the fibers in a unidirectional composite and parallel to the warp direction in a fabric-reinforced lamina. When the reinforcement is discontinuous fibers, whiskers or particulates, the choice of direction for the 1-axis is arbitrary, unless there is some preferred alignment direction for the reinforcement. Figure 1.3-1 defines the coordinate geometry for these materials.

Corresponding to these elastic properties is a set of parameters which commonly are used to characterize the strength of the material. There are many symbols used to denote these properties. We adopt the set employed in the U.S. Air Force Advanced Composites Design Guide and another set which uses subscripts keyed to the coordinate axes. The latter group is convenient for use with through-the-thickness properties, which are discussed later. We also include a notation (X_i^j, S_6) used in some strength analyses.

$F_1^{tu} = F_L^{ty}$ = Axial (Longitudinal) Tensile "Yield" Stress

$F_2^{tu} = F_T^{tu} = X_2^T$ = Transverse (Longitudinal) Tensile Ultimate Stress

$F_1^{cy} = F_L^{cy}$ = Axial (Longitudinal) Compressive "Yield" Stress

$F_1^{cu} = F_L^{cu} = X_1^c$ = Axial (Longitudinal) Compressive Ultimate Stress

$F_2^{ty} = F_T^{ty}$ = Transverse Tensile "Yield" Stress

$F_2^{tu} = F_T^{tu} = X_2^T$ = Transverse Tensile Ultimate Stress

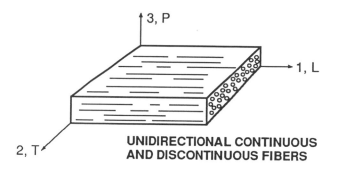

UNIDIRECTIONAL CONTINUOUS AND DISCONTINUOUS FIBERS

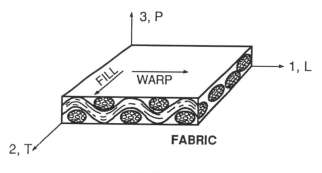

FABRIC

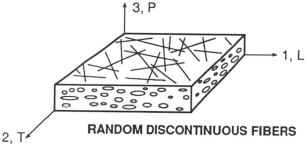

RANDOM DISCONTINUOUS FIBERS

FIGURE 1.3-1.

$F_2^{cy} = F_T^{cy}$ = Transverse Compressive "Yield" Stress

$F_2^{cu} = F_T^{cu} = X_2^c$ = Transverse Compressive Ultimate Stress

$F_{12}^{sy} = F_{LT}^{sy}$ = In-Plane Shear "Yield" Stress = Shear "Yield" Stress in the 1-2 Plane

$F_{12}^{su} = F_{LT}^{su} = S_6$ = In-Plane Shear Ultimate Stress = Shear Ultimate Stress in the 1-2 Plane

$F_{13}^{su} = F^{isu} = S_5$ = Interlaminar Shear Ultimate Stress = Shear Ultimate Stress in the 1-3 Plane

The quantity F^{isu} is the interlaminar shear ultimate stress in the 1-3 plane. There is another interlaminar shear ultimate stress in the 2-3 plane, a value rarely determined, which is discussed below.

Table 1.3-1a. Approximate values for through-the-thickness properties of composites based on in-plane properties of unidirectional composites having a similar fiber volume fraction.

Reinforcement	Property	Related Unidirectional Composite In-Plane Property	Estimated Percentage
Aligned Continuous or Discontinuous Fibers	Normal Extensional Modulus, E_3	Transverse Extensional Modulus, $E_2 = E_T$	60–100
	Shear Modulus in the 1–3 Plane, G_{13}	In-Plane Shear Modulus, $G_{12} = G_{LT}$	50–100
	Shear Modulus in the 2–3 Plane, G_{23}	In-Plane Shear Modulus, $G_{12} = G_{LT}$	40–100
	Normal Tensile Ultimate Stress, F_3^{tu}	Transverse Tensile Ultimate Stress, $F_2^{tu} = F_T^{tu}$	80–100
	Normal Compressive Ultimate Stress, F_3^{cu}	Transverse Compressive Ultimate Stress, $F_2^{cu} = F_T^{cu}$	80–100
	Shear Ultimate Stress in the 1–3 Plane, $F_{13}^{su} = F^{isu}$	In-Plane Shear Ultimate Stress, $F_{12}^{su} = F_{LT}^{cu}$	80–100
	Shear Ultimate Stress in the 2–3 Plane, F_{23}^{su}	In-Plane Shear Ultimate Stress, $F_{12}^{su} = F_{LT}^{su}$	50–100
	Poisson's Ratio in the 1–3 Plane, ν_{13}	Poisson's Ratio in the 1–2 Plane, ν_{12}	100
	Poisson's Ratio in the 2–3 Plane, ν_{23}	None	Absolute Value 0.25–0.32

Table 1.3-1b. Approximate values for through-the-thickness properties of composites based on in-plane properties of unidirectional composites having a similar fiber volume fraction.

Reinforcement	Property	Related Unidirectional Composite In-Plane Property	Estimated Percentage
Balanced Fabric	Normal Extensional Modulus, E_3	Transverse Extensional Modulus, $E_2 = E_T$	60–100
	Shear Modulus in the 1–3 Plane, G_{13}	In-Plane Shear Modulus, $G_{12} = G_{LT}$	50–100
	Shear Modulus in the 2–3 Plane, G_{23}	In-Plane Shear Modulus, $G_{12} = G_{LT}$	50–100
	Normal Tensile Ultimate Stress, F_3^{tu}	Transverse Tensile Ultimate Stress, $F_2^{tu} = F_T^{tu}$	25–90
	Normal Compressive Ultimate Stress, F_3^{cu}	Transverse Compressive Ultimate Stress, $F_2^{cu} = F_T^{cu}$	90–200
	Shear Ultimate Stress in the 1–3 Plane, $F_{13}^{su} = F^{isu}$	In-Plane Shear Ultimate Stress, $F_{12}^{su} = F_{LT}^{su}$	50–90
	Shear Ultimate Stress in the 2–3 Plane, F_{23}^{su}	In-Plane Shear Ultimate Stress, $F_{12}^{su} = F_{LT}^{su}$	50–90
	Poisson's Ratio in the 1–3 Plane, ν_{13}	Poisson's Ratio in the 1–2 Plane, ν_{12}	100
	Poisson's Ratio in the 2–3 Plane, ν_{23}	Poisson's Ratio in the 1–2 Plane, ν_{12}	100

Table 1.3-1c. Approximate values for through-the-thickness properties of composites based on in-plane properties of unidirectional composites having a similar fiber volume fraction.

Reinforcement	Property	Related Unidirectional Composite In-Plane Property	Estimated Percentage
Randomly Oriented Fibers in the 1–2 Plane	Normal Extensional Modulus, E_3	Transverse Extensional Modulus, $E_2 = E_T$	60–100
	Shear Modulus in the 1–3 Plane, G_{13}	In-Plane Shear Modulus, $G_{12} = G_{LT}$	60–100
	Shear Modulus in the 2–3 Plane, G_{23}	In-Plane Shear Modulus, $G_{12} = G_{LT}$	60–100
	Normal Tensile Ultimate Stress, F_3^{tu}	Transverse Tensile Ultimate Stress, $F_2^{tu} = F_T^{tu}$	40–100
	Normal Compressive Ultimate Stress, F_3^{cu}	Transverse Compressive Ultimate Stress, $F_2^{cu} = F_T^{cu}$	70–100
	Shear Ultimate Stress in the 1–3 Plane, $F_{13}^{su} = F^{isu}$	In-Plane Shear Ultimate Stress, $F_{12}^{su} = F_{LT}^{su}$	50–90
	Shear Ultimate Stress in the 2–3 Plane, F_{23}^{su}	In-Plane Shear Ultimate Stress, $F_{12}^{su} = F_{LT}^{su}$	50–90
	Poisson's Ratio in the 1–3 Plane, ν_{13}	Poisson's Ratio in the 1–2 Plane, ν_{12}	100
	Poisson's Ratio in the 2–3 Plane, ν_{23}	Poisson's Ratio in the 1–2 Plane, ν_{12}	100

For most structures, a knowledge of the in-plane strength and elastic properties is all that is required for adequate design; however, in a few instances, through-the-thickness effects may be important. For example, the normal (through-the-thickness) extensional moduli of composites are relatively low, and large strains in the normal direction can lead to deformations which affect the in-plane stress distribution in structures such as pressure vessels and flywheels. Similarly, the two through-the-thickness shear moduli G_{13} and G_{23} are also relatively small, and the assumption that the corresponding shear strains γ_{13} and γ_{23} are negligible may not be valid in some cases. Normal extensional and shear stresses can also be very important in regions where the ratio of thickness to radius of curvature is small and near joints. These subjects are covered in Volume 5 on design. Our concern here is with the elastic and strength properties required in these situations. The basic through-the-thickness properties are (see Figure 1.3-1):

E_3 = Normal Extensional Modulus
G_{13} = Shear Modulus in the 1-3 Plane
G_{23} = Shear Modulus in the 2-3 Plane
ν_{13} = Poisson's Ratio in the 1-3 Plane[4]

[4]Definition follows form of Equation (1.3-1).

ν_{23} = Poisson's Ratio in the 2-3 Plane[4]
F_3^{tu} = Normal Tensile Ultimate Stress
F_3^{cu} = Normal Compressive Ultimate Stress
F_{23}^{su} = Shear Ultimate Stress in the 2-3 Plane

Finding reliable data for through-the-thickness properties is particularly difficult. If they are not available, it is possible to obtain rough estimates for these quantities, which are good enough for preliminary design purposes, as a percentage of in-plane properties of unidirectional composites having a similar fiber volume fraction. These estimates are convenient because frequently there are more data available for unidirectional composites than for materials with other forms of reinforcement. Tables 1.3-1a–c list approximations for composites reinforced with aligned continuous and discontinuous fibers, fabrics and random discontinuous fibers. The estimated values should agree with actual properties to within about 50%. This error range may seem excessive, but, as we shall see, it is no worse than the variation in composite properties which may be encountered when comparing data for the same material from different sources. For final design, more accurate property values should be obtained by direct measurement using materials fabricated by the final manufacturing method, or one close to it. Note that since fibers used as reinforcements invariably are stiffer than resins,

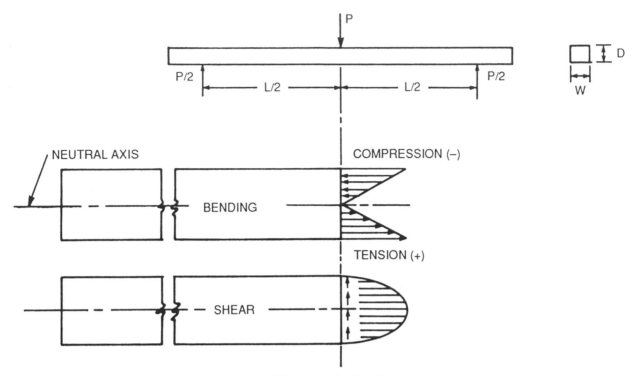

FIGURE 1.3-2. Three-point loading flexure test.

all composite shear and extensional moduli should be greater than the corresponding matrix properties. Typical extensional modulus and shear modulus values for thermosetting resins, such as epoxies, vinyl esters and polyesters, are 3.5 GPa (5×10^5 psi) and 1.4 GPa (2×10^5 psi), respectively.

The quantities defined above are commonly accepted as the basic elastic and strength properties of composites. Flexural modulus and strength, although frequently reported in the literature and in manufacturers' data sheets, are *not* intrinsic properties. The three-point and four-point flexure tests subject specimens to tensile, compressive and shear stresses [3]. Consequently, measured deflections and failure loads reflect this complex state of stress. The subject of test methods is covered in Volume 6, and we will not consider it here. However, the significance of flexural test results, since they are widely reported, deserve further consideration.

Consider a beam specimen for which the reinforcement is uniformly distributed throughout the material. We assume that the beam axis is parallel to one of the material principal axes. Assume the specimen is subject to three-point loading. Figure 1.3-2 shows that the state of stress at the mid-beam cross section, according to

simple Bernoulli-Euler beam theory, consists of a linearly varying flexural stress, with tension on the bottom surface and compression on the top, and a parabolic shear stress distribution. We are interested in the flexural modulus and strength of such a specimen. Consider modulus first.

As discussed by Zweben, Smith and Wardle [4], both shear deformation and flexural deformation contribute to the deflection at the center of the beam. The very name flexural modulus implies a material property associated with deflection of a structure due to *bending* and independent of shear. Therefore, to obtain a true measure of flexural modulus, specimen geometry should be chosen such that the influence of shear deformation is negligible. As explained in reference [4], the ratio of shear deflection to bending deflection decreases as the beam aspect ratio (ratio of length to thickness), *L/D*, increases. It is important to realize that, *when the tensile and compressive moduli of a material are equal, and the reinforcement is uniformly distributed, the flexural modulus should equal the tensile and compressive moduli*. Figure 1.3-3, taken from the reference cited above, shows that the apparent flexural modulus does approach the measured tensile modulus as the aspect ratio increases. At an *L/D* ratio of about 50, the two values are approximately equal. (The relatively small difference can be attributed to experimental uncertainty.) As discussed in reference [4], the aspect ratio required to obtain valid flexural moduli depends on the ratio of extensional modulus-to-transverse shear modulus, E_1/G_{13}.

Note that, although beams with fairly large aspect ratios are required for modulus, large deflections can occur which require corrections for calculation of ultimate strength [4]. As Figure 1.3-2 shows, tensile, compressive and shear stresses all exist in a flexure specimen. Consequently, failure can occur by mechanisms associated with any of these stresses acting individually or in combination. In fact, the stress at failure in a flexural test may be higher than the material's tensile, compression and shear strengths. This subject is discussed further later on.

Variability

The primary purpose of this part is to present the designer with representative elastic and static strength properties for use in material trade-off studies and preliminary designs. At first glance, this would seem like a fairly straightforward exercise which is not the case. In fact, property data for a given material reported by different sources often vary greatly. Further, individual organizations often observe considerable variation in the mechanical properties of a given system. In short, at the present time we cannot speak of "the" properties of a particular fiber/matrix combination, and the question of material variability requires careful consideration. In this discussion, we consider the nature of property variability, its sources, consequences, and methods of establishing allowable properties for design.

There are several kinds of property variability that must be considered, among which are:

- inherent
- batch-to-batch
- point-to-point within a material
- unintentional variation of properties with direction at a point
- spurious data from invalid test methods or poorly conducted tests

Some of the sources of composite property variability are:

- inherent and production-related fiber and matrix property variability

- variations in intermediate materials (e.g., prepregs, sheet molding compound)
- variations in fabrication processes
- local and overall (global) variations in fiber volume fraction
- variations in fiber orientation resulting from various sources such as resin flow and poor initial placement
- voids

First, we consider some aspects of test methods that relate to variability. There are many different tests used throughout the industry to measure a given property. Even tests for which standard methods exist cannot always be relied upon to give valid data. For example, the widely used flexural test method, ASTM D790-71, recommends beam aspect ratios of 16:1, 32:1, and 40:1. As Figure 1.3-3 shows, each of these ratios gives rise to a different value for flexural modulus of some materials because of the shear deformation effect described earlier.

To illustrate how the use of different tests contributes to the variation in reported properties, consider Table 1.3-2 which presents data for the shear modulus of unidirectional graphite/epoxy obtained by Yeow and Brinson [5] using some of the most common methods. The difference between the largest and smallest modulus

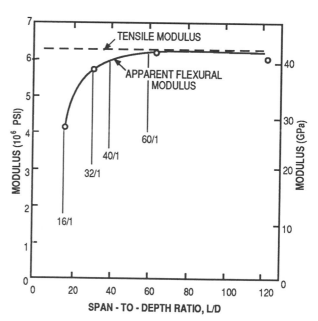

FIGURE 1.3-3. Apparent flexural modulus as a function of span-to-depth ratio for a composite with unidirectional Kevlar 49 aramid fibers in a polyester resin matrix.

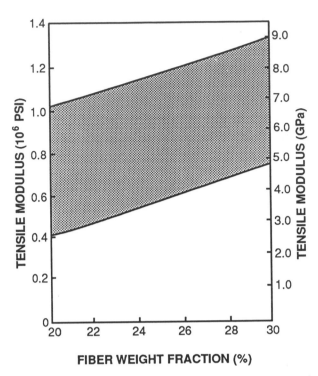

FIGURE 1.3-4. Tensile modulus of *E* glass mat/polyester hand lay-up laminates.

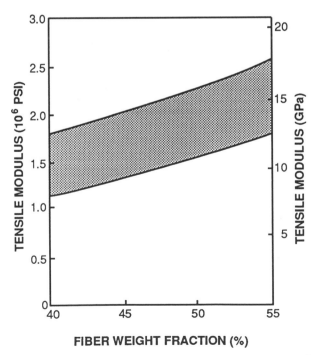

FIGURE 1.3-5. Tensile modulus of *E* glass woven roving/ polyester hand lay-up laminates.

Table 1.3-2. Comparison of shear properties from different test methods.

Method	Test Laminate Geometry	Initial Modulus (GPa)	Ultimate Stress (MPa)	Ultimate Strain (%)
Symmetric Rail Shear	$[0]_{8s}$	6.90	—	—
Symmetric Rail Shear	$[0/90]_{4s}$	4.07	>90	>8
Off-Axis Tension	$[10]_{8s}$	7.79	85	2.0
Off-Axis Tension	$[15]_{8s}$	8.41	68	1.6
[±45] Tension— Rosen	$[\pm45]_{4s}$	5.17	~90	~4
[±45] Tension— Petit	$[\pm45]_{4s}$	4.62	~90	~4

values is over 100%, and ultimate strains differ by as much as a factor of more than five. Because test method development and standardization have proceeded slowly, this situation is likely to persist.

Although testing problems may account for some of the differences in published property data, there is a real and inherent variability in even the most carefully fabricated materials [6]. This variability can be attributed to the basic scatter in fiber and matrix properties, fiber spatial distribution, and void content. Section 1.2 discusses these factors in greater detail.

To illustrate the magnitude and importance of variability, we consider examples of composites reinforced with random, unidirectional and discontinuous fibers or woven fabrics which are fabricated by different methods. Reference [7] summarizes data from an industry-wide survey of properties of hand lay-up *E* glass/polyester composites made with mat and woven roving reinforcements. The following figures show data from that study. Figures 1.3-4 and 1.3-5 present results for tensile modulus of mat and woven roving laminates, respectively, as functions of glass weight content. Data for laminates containing woven rovings were measured in the warp direction. We note that there is a considerable range of observed glass contents. For example, measured fiber weight fractions for mat (which can be related to fiber volume fraction, V_f, using methods discussed in section 1.2) ranged from 20 to 30 percent. Differences in V_f generally are an important source of the variation in published properties. Often, fiber volume fraction is not reported at all, or an estimated or nominal value is used. However, even at a fixed value of fiber content, there is considerable property variability. For example, at a fiber weight content of 20%, mat tensile modulus ranges from 2.8 GPa (0.4 × 10⁶ psi) to over 6.9 GPa (1 × 10⁶ psi). The upper limit is 150% of the lower limit, a remarkable difference. Without further information, it is not possible to determine whether the reported variation results from real property variability or poor test methods. We suspect that both factors contribute.

Figure 1.3-6 presents data [7] for flexural strength of mat laminates as a function of fiber weight fraction. We observe that, as in the case of modulus, there is considerable property variability. Flexural strength is a response to a complex loading. It is not a basic material property as are tensile strength, compressive strength and shear strength.

Figure 1.3-7 presents data for the tensile strength of *E* glass woven roving/polyester hand lay-up laminates as a function of fiber weight fraction. Again, we observe substantial property variability which is also evident in the compressive strength data of Figure 1.3-8. To emphasize the point that flexural strength levels usu-

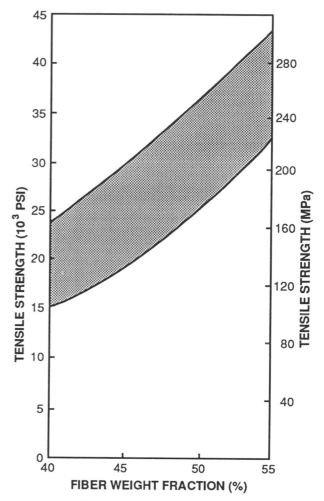

FIGURE 1.3-7. Tensile strength of *E* glass woven roving/ polyester hand lay-up laminates.

ally are different from those of either tension or compression, consider Figure 1.3-9 which shows how flexural strength varies with glass content. There is not a one-to-one correspondence among the three strength values. The range of flexural strength values is generally above those of tensile and compressive strengths.

The considerable variability evident in Figures 1.3-4 to 1.3-9 probably can be attributed to a great extent to real differences in properties, although some of the scatter undoubtedly results from use of different test methods by the organizations from which the data were obtained. Hand lay-up fabrication methods, which were used for the materials discussed above, are among the least controllable; thus, significant variability is to be expected.

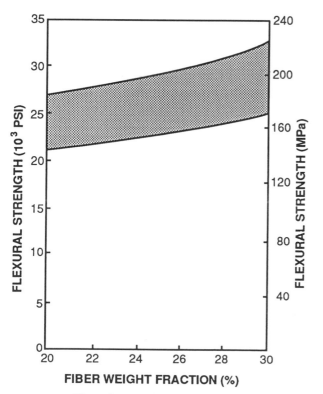

FIGURE 1.3-6. Flexural strength of *E* glass mat/polyester hand lay-up laminates.

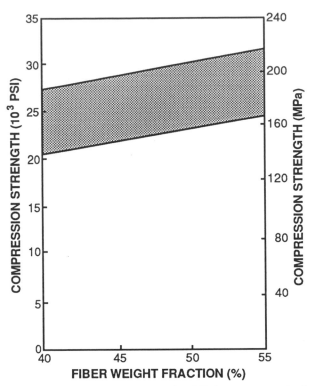

FIGURE 1.3-8. Compressive strength of *E* glass woven roving/ polyester hand lay-up laminates.

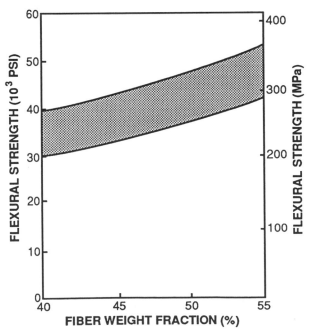

FIGURE 1.3-9. Flexural strength of *E* glass woven roving/ polyester hand lay-up laminates — warp direction.

Variation in reported composites properties is not unique to the marine industry or the hand lay-up process. Table 1.3-3 presents mechanical property data for a widely used unidirectional graphite/epoxy system reported by manufacturers in five government contract reports selected at random. The last column shows the difference between the highest and lowest values. Data of sources 1 to 3 are mean values. Moduli in column 4 are mean values, but strengths are 80% of measured averages. Data in column 5 are design allowables which are lower than mean values to account for scatter. The subject of design allowables is considered later.

We see that, even accounting for the different bases used to compute the tabulated data, there is considerable variability. For example, the lowest value of compressive strength (Company 2) is 868 MPa (126 × 10³ psi), while the highest (Company 1) is 1930 MPa (280 × 10³ psi), 122% greater. Both are mean values. There are also substantial differences in compressive modulus. It is noteworthy that there is little correlation between columns 2 and 3 which represent data from two divisions of the same company.

These data obviously raise some significant questions, the main one being whether the property variability is real or simply the result of poor test methods. The information presented in the reports is not sufficient to make such a determination possible. In fact, the test methods used frequently are not reported. Some of the differences may result from differences in fiber volume fractions (which are not reported for most of the data in the table) and perhaps inexperience in processing the system. Much of the variability is probably a reflection of poor test procedures or use of different test methods. We would expect this variability for compressive strength in particular because it is one of the more difficult properties to measure [8].

The property variability discussed above may be somewhat dismaying to the engineer encountering them for the first time; however, it should be kept in perspective. Reliable fiberglass boats are produced by the tens of thousands annually using the hand lay-up fabrication method. At the other end of the technological spectrum, high-performance structural components made from boron, graphite, and aramid fiber composites have been used for several years throughout the aerospace industry, and the number of applications for these materials is increasing. The success of the latter structures is based on extensive developmental testing to ensure reliable performance under the loading and en-

Table 1.3-3. Variation of reported unidirectional properties for a widely used graphite/epoxy system (Sources: Major Airframe Company Reports).

Property	Source					Maximum Difference (%)
	1	2*	3*	4	5	
Elastic Constants (10⁶ psi)						
Longitudinal Tensile Modulus	20.8	18.1	21	20.6	18.5	16
Longitudinal Compressive Modulus	18.6	14.5	21	19.8	18.5	45
Transverse Tensile Modulus	1.9	1.8	1.7	1.3	1.6	46
In-Plane Shear Modulus	0.85	–	0.65	0.8	0.65	31
Poisson's Ratio (Dimensionless)	0.30	–	–	0.32	0.25	28
Strength Properties (10³ psi)						
Longitudinal Tension	274	190	180	164	169	67
Longitudinal Compression	280	126	180	126	162	122
Transverse Tension	9.5	5.2	8	5.4	6.0	83
Transverse Compression	39	–	30	21	25	86
In-Plane Shear	17.3	–	12	8.4	–	106
Interlaminar Shear	–	13.5	13	–	7.1	90

*Divisions of the same company.

vironmental conditions for which they were designed. The ability to make durable fiberglass boats has been primarily an empirical trial and error process. In both industries, continued production of reliable structures depends strongly on careful control of raw material quality and fabrication processes.

Design Allowables

Generally speaking, there are two approaches to the problem of composite property variability. The first method uses fixed, deterministic design allowables which are lower than mean values to account for scatter. The second design approach considers mechanical properties, and usually loads as well, to be statistical variables and works with probabilities of failure. For a discussion of this method, see references [9–14].

In this discussion we deal primarily with the deterministic approach because statistical design methods are extremely complex and it is extremely rare to find enough material property data to establish a reasonable cumulative distribution function.

Even though few designers use a probabilistic design approach, it is widely recognized that when there is significant scatter in material properties, mean values can-

not be used. Instead, two types of reduced values commonly are used: "A" basis design allowables and "B" basis design allowables. The following definitions of these quantities are taken from MIL Handbook 5 [15]:

- *"A" Basis*—The A value is the value above which at least 99% of the population of values is expected to fall, with a confidence of 95%.
- *"B" Basis*—The B value is the value above which at least 90% of the population of values is expected to fall, with a confidence of 95%.

A third type of allowable, the "S" basis value, is used when a minimum property level can be assured. This is rarely the case for composites, unless some sort of proof test is used.

The "A" and "B" basis allowables depend not only on the amount of scatter, but also on the number of specimens used. Procedures for evaluating these allowables can be found in MIL Handbook 5 or any standard text on statistics.

"A" basis design allowables are lower than "B" values and are used for critical components. "B" allowables are generally used for secondary structures, failure of which would not cause harm to life or severe property damage.

Table 1.3-4. Comparison of design allowable stresses for aluminum and for unidirectional graphite/epoxy at room temperature.

Material	Strength Property	Allowable Stress (MPa)			Coefficient of Variation (%)
		Mean	"B" Value	"A" Value	
Aluminum[1]	Tensile Yield	–	280	270	–
(2024-T3)	Tensile Ultimate	–	450	450	–
Graphite/Epoxy[2]	Longitudinal Tension	1539	1171	914	10.3
	Longitudinal Compression	1393	1000	727	11.7
	Transverse Tension	42.3	34.6	29.3	7.8
	Transverse Compression	147.6	101.2	68.8	13.5
	In-Plane Shear	95.4	86.7	80.7	3.9

[1]Reference [15].
[2]Reference [16].

The differences between mean values and "B" allowables, and between "B" allowables and "A" allowables increase as property variability increases. To get an idea of the difference in the extent of mechanical property variability between composites and typical structural metals, consider Table 1.3-4 which presents MIL Handbook 5 [15] design allowables for 2024-T3 aluminum and for a typical unidirectional graphite/epoxy system, from the work of G. M. Lehman et al. [16]. We see that while the aluminum "A" and "B" allowables differ by only a few percent, the composite "A" allowables are as much as 32% lower than the corresponding "B" allowables. We also note that the composite "A" basis design allowables are substantially less than the respective mean properties. As the composite design allowables were based on several hundred tests, the reductions with respect to mean values can be attributed primarily to data scatter, rather than use of a small number of specimens.

Table 1.3-4 also presents coefficients of variation for the graphite/epoxy strength data. Coefficient of variation, which is defined as standard deviation divided by mean, is a common measure of scatter. Values for the particular system studied varied between 3.9 and 13.5%. Generally, scatter in modulus data is significantly lower than that for strength properties. The minimum coefficient of variation for composite mechanical properties that can be expected is about 2½ to 3%.

We should point out that, even though the composite design allowables are substantially lower than mean values, use of composites resulted in substantial weight savings for the structural components developed in the program which generated these data.

In establishing design allowables, it is important that the test specimens used accurately reflect the processes by which the actual structure will be made, including possible batch-to-batch variations. To assure that the properties that exist in the structure are consistent with the design allowables, quality assurance procedures must be established. This assurance includes acceptance test procedures for incoming materials, methods to assure that raw materials in stock retain acceptable property levels, and careful controls on processing.

Size Effects

Another potentially important aspect of variability that may need to be considered in design of very large structures is the "size effect." This effect refers to the decrease of mean strength with increasing material volume that is observed when flaw-sensitive materials like brittle ceramics are tested [17–21]. The qualifying words "potentially" and "may" are used because size effect has not been demonstrated definitively in composites, although there are some experimental data to support its existence and analytical models for composite tensile strength predict such a phenomenon [22–28].

The physical explanation for the size effect in defect-sensitive materials is that defect intensity is a statistical variable, and the probability of finding a serious flaw increases as material volume increases. Since composite strength properties display considerable scatter, a characteristic which is also common to most brittle materials, there is a strong suspicion that they are defect-sensitive materials and, therefore, should have volume-dependent strength characteristics.

Riedinger, Kural and Reed [19] collected data on the tensile strength of glass fibers, test coupons, small pressure vessels, and large rocket motor cases. Figure 1.3-10 shows that the results of their study indicate a significant downward trend in strength as volume increases. Although these data suggest the existence of a size effect, they are not conclusive for several reasons. First, laboratory coupons and fiber test samples are loaded in pure tension, but pressure vessels are subjected to complex, nonuniform stresses throughout the structure. Second, pressure vessel fabrication probably introduces manufacturing defects that are not present in laboratory test specimens. In either case, the data show that use of coupons to generate design data for large structures requires caution.

An indication for a size effect rests on the observation that for most inorganic fiber composites, flexural strengths are consistently higher than those of identical specimens tested in pure tension [29]. The argument here is that when coupons are tested in flexure, the volume of material subjected to high tensile stresses is limited to the small region near the outer surface of the beam near the center of the span, and this smaller highly loaded volume accounts for the higher strength.

The significance of a size effect for design is that material properties data usually are determined using relatively small test specimens. Large structures, such as chemical storage tanks, airplane wings and rocket motor cases, may have material volumes that are orders of magnitude greater. If a size effect exists, structural failure stress levels could be substantially lower than design allowables based on laboratory coupons.

For design purposes, there are some relatively simple procedures for estimating the magnitude of a possible size effect based on the work of Weibull [30,31]. He postulated a strength distribution function for flaw-sensitive materials of the form:

$$F(S) = 1 - \exp\left[-V(S/S_0)^m\right] \quad (1.3-3)$$

where S_0 is a reference strength level, m is a shape parameter which is inversely related to scatter, and V is the material volume which is subjected to the constant stress level S.

For a Weibull material [one whose strength is described by Equation (1.3-3)], the ratio of mean strength of two volumes, V_1 and V_2, is given by:

$$\overline{S}_2/\overline{S}_1 = (V_1/V_2)^{1/m} \quad (1.3-4)$$

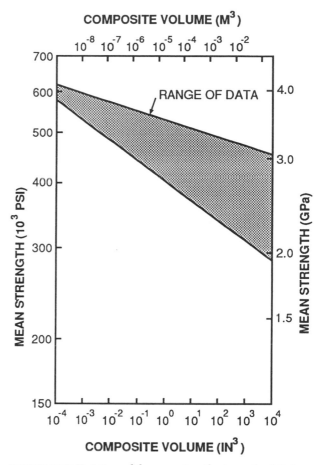

FIGURE 1.3-10. Variation of the mean tensile strength of S glass composites with volume [19].

The Weibull shape parameter, m, depends only on the coefficient of variation which we denote by c. Figure 1.3-11 shows how m varies with c. A simple approximate formula for estimating m that gives reasonably good agreement with the exact relationship which Figure 1.3-11 shows, is

$$m = \frac{1.2}{c} \quad \text{Approximate} \quad (1.3-5)$$

Figure 1.3-12 shows how the ratio of mean strengths, $\overline{S}_1/\overline{S}_2$, decreases as the volume ratio, V_2/V_1, increases for various values of m. The rate of decrease increases as material scatter increases (recall that m is inversely related to scatter, so that small values of m correspond to a large amount of scatter). We note that the predicted difference in mean strengths between large and small volumes can be quite significant, even for an m value of

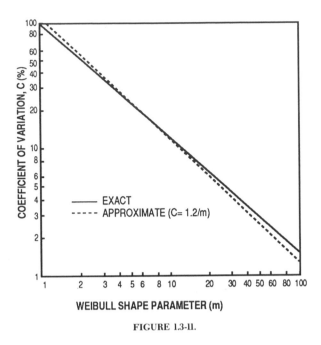

FIGURE 1.3-11.

strength values and better elevated temperature characteristics than polyesters or vinyl esters. However, considerations such as resin cost, toxicity factors, ease of fabrication, performance at very high temperatures, and environmental stability may dictate the use of other matrices. There have been extensive research programs involving use of thermoplastic resins with unidirectional composites, primarily polysulfones and polyetheretherketones, that may lead to improved products and processes. The potential fabrication cost advantages and damage tolerance of thermoplastics are strong driving forces for their use, and further work with these systems will provide guidelines for the designers of composite structures and their application [32,33].

As was indicated, epoxy-matrix composites tend to provide among the highest room temperature strengths, and there are more data available for unidirectionally reinforced materials using these matrices than for other resins. For these reasons, we have selected epoxy-matrix systems as baselines for unidirectional composites.

The apparent reason for the high strengths obtained with epoxy matrices is that they bond well to fibers, although it took considerable work on surface treatments before a good bond was made with graphite (carbon) fibers. We anticipate that, in due course, development of improved finishes and surface treatments will improve the strengths of composites using other thermoset resins. (We note that considerable work has been done on finishes for glass fibers for use in polyester and vinyl ester resins.) At the present time, the designer should exercise discretion in using strength values for epoxy composites for other resin systems. Elastic properties, however, are quite similar, as long as fiber volume fractions do not differ substantially.

Section 1.2 presents a fairly extensive discussion of the properties of unidirectional composites and how they relate to those of their constituents. In this section, we briefly discuss the static stress-strain characteristics of composites to provide the designer with some insight into the mechanical behavior of these materials on the macroscopic level.

50 which corresponds to a coefficient of variation of only 2.5%, about the minimum that can be expected in composites at their present stage of development.

Figure 1.3-12 provides a direct way to estimate the amount of strength reduction related to a possible size effect. It is important that the value of m used incorporate the influence of actual manufacturing defects on strength. This requirement may demand that coupons be cut from large structural elements made using the full-scale fabrication equipment. The possibility of a size effect places additional emphasis on the importance of subcomponent, component, and full-scale tests in the development of large structures.

1.3.3 Unidirectional Composites

Unidirectional composites, which contain straight, parallel, continuous fibers, are the most efficient form of fibrous material. For this reason, they have been widely studied throughout the aerospace industry and are probably the best characterized class of composite materials. To date, the major resins used have been mostly thermosets, epoxies, polyesters, vinyl esters and polyimides. Epoxies have been the major type of resin used in aerospace applications; as a result, the bulk of available data has been obtained for composites employing these polymers as matrices. So far, epoxies appear to provide the highest room temperature

Stress-Strain Properties of Unidirectional Materials

Generally, the stress-strain curves of unidirectional composites are linear to failure, a characteristic they share with brittle materials. Let us consider this subject in greater detail.

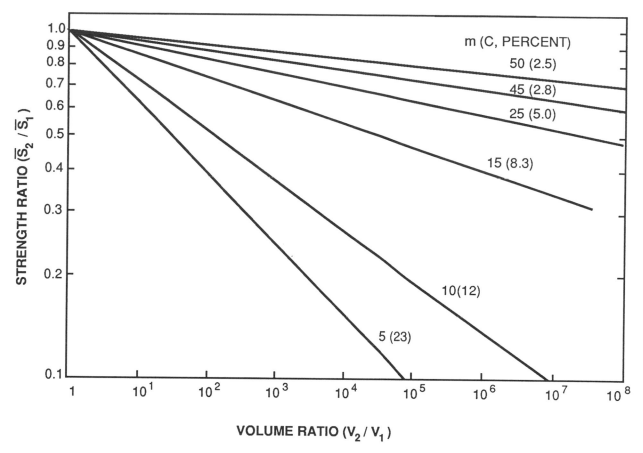

FIGURE 1.3-12.

The longitudinal tensile stress-strain curves of almost all composites are linear, except perhaps at very high stress levels where the occurrence of a large number of scattered fiber breaks can cause a reduction of modulus. This departure from linearity is of little or no practical significance. The major exceptions to the tensile linearity rule are composites reinforced with Kevlar 29 aramid fibers. These fibers display a significant increase in modulus with increasing tensile stress. We note, however, that Kevlar 49 fibers, which are more commonly used in composites, have linear tensile stress-strain curves.

Occasionally, fabrication methods introduce unintended fiber curvature which results in a nonlinear tensile stress-strain curve having a slope that increases with increasing stress. Apparently, the increasing load straightens the fibers, resulting in a higher tensile modulus [34,35]. Zweben [34] examined unidirectional composites made from prepreg with "wavy" Kevlar 49 aramid fibers. Figure 1.3-13 shows the tensile stress-

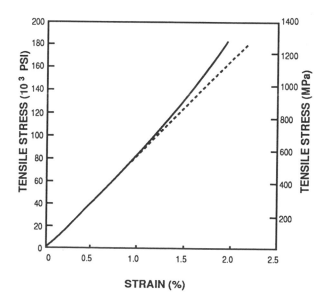

FIGURE 1.3-13. Tensile stress-strain curve for a unidirectional aramid fiber composite with initially "wavy" fibers [34].

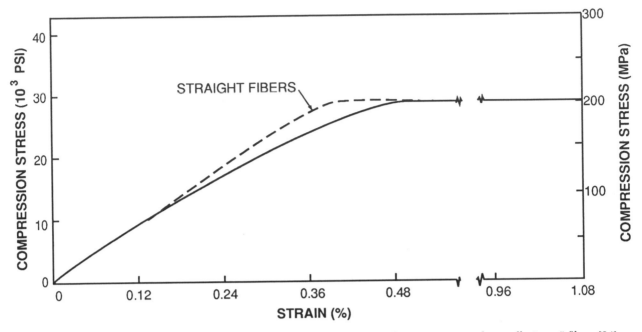

FIGURE 1.3-14. Compressive stress-strain curve for a unidirectional aramid fiber composite with initially "wavy" fibers [34].

strain curve for these specimens. Another major source of fiber curvature is resin flow which can occur in many fabrication processes. Significant fiber curvature in unidirectional composites is undesirable because it causes local reduction in stiffness which can result in internal stress redistribution and possibly component failure. Curvature can also adversely affect local strength properties.

The axial compressive and tensile moduli of unidirectional composites usually are considered to be equal. However, graphite composites often are reported as having a lower compressive modulus, as Table 1.3-3 shows. It is not clear at this time whether this anomaly results from real differences in fiber tensile and compressive moduli, fiber curvature, faulty test methods, or some other undetermined factors. Generally, the reported difference ranges between about 0 and 15%. The compressive modulus in Table 1.3-3, Column 2 is 20% lower than the tensile modulus. Once in a while, a publication is encountered which reports a higher value for compressive modulus, but this is less common.

As design and analysis of structures made from materials having different tensile and compressive moduli (sometimes called bi-modulus materials) is extremely complex [36], a single value is almost always used. In most situations, adopting the lower value would be conservative.

Most unidirectional composites have compressive stress-strain curves which are linear to failure. The major exceptions are materials with curved fibers and aramid fiber composites, both of which will be considered shortly. Fiber curvature can cause nonlinear composite compressive stress-strain curve, even when the fibers themselves behave linearly under compressive loading. However, whereas curved fibers result in a composite tensile modulus that increases with increasing stress, fiber curvature causes compressive modulus to decrease with applied stress.

Kevlar 49 aramid fibers have a nonlinear compressive stress-strain curve, and this is reflected in composite behavior, as reported by Greenwood and Rose [37] and Kulkarni, Rice and Rosen [38]. Figure 1.3-14, which is taken from the work of Zweben [34], illustrates both the nonlinear compressive stress-strain characteristics of composites made with Kevlar 49 fibers and the additional nonlinearity resulting from fiber curvature. The compressive specimens used were made from the same prepreg as we used for the tensile specimens of Figure 1.3-13, which has "wavy" fibers. The compressive stress-strain curve of a composite with straight Kevlar 49 fibers would be expected to follow the dashed line shown in Figure 1.3-14. The shape of this curve is reminiscent of that of an elastic, perfectly plastic material. However, there is no plastic deformation involved.

The actual curve departs from the dashed line, apparently as a result of fiber curvature.

The transverse tensile stress-strain curve of unidirectional composites is generally linear to failure. The transverse tensile ultimate strain is usually quite small, less than 0.5%, as a rule. The transverse compressive ultimate strain is typically much greater than this, and the compressive ultimate strain curve frequently displays some nonlinearity at the upper end of the curve.

The in-plane shear stress-strain curve of unidirectional composites is almost always nonlinear. Figure 1.3-15, based on the work of Slepetz, Zagaeski and Novello [39] presents a typical result for high-tensile strength graphite epoxy. We note that there is a substantial difference between the initial tangent modulus and the secant modulus drawn to the failure point. The importance of this nonlinear behavior for design depends on the orientations of the layers and the loading conditions and must be evaluated on a case-by-case basis.

Representative Properties of Unidirectional Composites

This discussion gives representative room temperature mechanical properties of the most important types of unidirectional composites for use in preliminary design and trade-off studies. The fibers included are *E* glass, *S* glass, aramid (Kevlar 49), high-strength graphite, high-modulus graphite, ultrahigh-modulus graphite, boron, and alumina. As discussed earlier, strength properties are typical of epoxy-matrix systems and those of other matrices may be somewhat lower.

We have elected to present properties of generic types of fibers, such as high-strength graphite, rather than one particular product for several reasons. One is that fiber technology is in a continuing stage of development; fiber constituents, precursors and manufacturing processes can be expected to change to improve properties, to reduce raw materials and processing costs, and to increase plant productivity. For example, the formulation of *E* glass manufactured by a leading producer was changed within the last few years, increasing the specific gravity from 2.54 to 2.60. According to the manufacturer, there is little effect on properties, but the designer cannot assume that properties will always remain unchanged for all fibers.

Another reason for selecting generic classes of fibers is that for some categories, particularly high-strength graphite and *E* glass, there are many products with similar properties. Further, over the years some products have disappeared from the market and new ones have appeared.

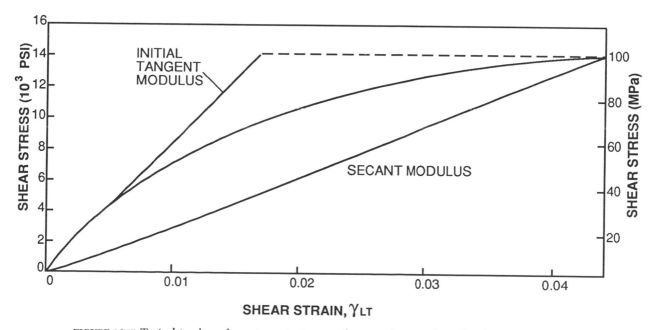

FIGURE 1.3-15. Typical in-plane shear stress-strain curve for a unidirectional graphite/epoxy composite [39].

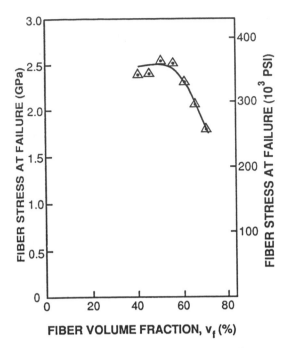

FIGURE 1.3-16. Fiber tensile stress at failure as a function of fiber volume fraction for E glass/epoxy strands [40].

Table 1.3-5. *Representative properties of* E *glass/epoxy unidirectional composites: fiber volume fraction,* $v_f = 0.60.$

	GPa	10^6 psi
Elastic Constants		
Longitudinal Modulus, E_L	45	6.5
Transverse Modulus, E_T	12	1.8
Axial Shear Modulus, G_{LT}	5.5	0.8
Poisson's Ratio, ν_{LT} (dimensionless)	0.28	
Strength Properties	**MPa**	10^3 psi
Longitudinal Tension, F_L^{tu}	1020	150
Longitudinal Compression, F_L^{cu}	620	90
Transverse Tension, F_T^{tu}	40	7
Transverse Compression, F_T^{cu}	140	20
In-Plane Shear, F_L^{su}	70	10
Interlaminar Shear, F^{isu}	70	10
Ultimate Strains (Percent)		
Longitudinal Tension, ϵ_L^{tu}	2.3	
Longitudinal Compression, ϵ_L^{cu}	1.4	
Transverse Tension, ϵ_T^{tu}	0.4	
Transverse Compression, ϵ_T^{cu}	1.1	
In-Plane Shear, γ_{LT}^u	1–6	
Density, kg/m³ (lb/in³)	2.1×10^3 (0.075)	

Table 1.3-6. *Representative properties of* S *glass/epoxy unidirectional composites: fiber volume fraction,* $v_f = 0.60.$

	GPa	10^6 psi
Elastic Constants		
Longitudinal Modulus, E_L	55	8.0
Transverse Modulus, E_T	16	2.3
Axial Shear Modulus, G_{LT}	7.6	1.1
Poisson's Ratio, ν_{LT} (dimensionless)	0.28	
Strength Properties	**MPa**	10^3 psi
Longitudinal Tension, F_L^{tu}	1620	230
Longitudinal Compression, F_L^{cu}	690	100
Transverse Tension, F_T^{tu}	40	7
Transverse Compression, F_T^{cu}	140	20
In-Plane Shear, F_L^{su}	80	12
Interlaminar Shear, F^{isu}	80	12
Ultimate Strains (Percent)		
Longitudinal Tension, ϵ_L^{tu}	2.9	
Longitudinal Compression, ϵ_L^{cu}	1.3	
Transverse Tension, ϵ_T^{tu}	0.4	
Transverse Compression, ϵ_T^{cu}	1.1	
In-Plane Shear, γ_{LT}^u	1–6	
Density, kg/m³ (lb/in³)	2.0×10^3 (0.073)	

The designer cannot obtain composite properties from a handbook with the same degree of confidence as for metallic alloys. This fact is graphically illustrated by the variability in the data for a particular graphite/epoxy system presented in Table 1.3-1. This inconsistency presented considerable difficulties in assembling tables of properties for this section. What has been done is to examine data from many sources and construct what is believed to be a reasonably consistent set of representative data. The reputation of the sources was evaluated and some of the theoretical considerations of section 1.2 were used to screen out obviously inconsistent data.

One of the simplest and most effective ways of judging the validity of data when fiber volume fraction is presented is to compare the composite tensile modulus with the rule of mixtures prediction (see subsection 1.2.2).

$$E_L = E_1 = v_f E_f + v_m E_m \qquad (1.3\text{-}6)$$

where E_f is fiber extensional modulus, E_m is matrix extensional and v_f and v_m are fiber and matrix volume fractions, respectively. Typical fiber and matrix properties are presented in section 1.1.

As indicated in section 1.2 and shown in Equation

(1.3-6), composite properties depend strongly on fiber volume fraction, v_f. This quantity typically ranges from 50 to 70% for unidirectional composites, depending on the material system and manufacturing process. As a reference level, we have selected 60%, which is in the middle of the spread of values.

There are some simple procedures that can be used to obtain estimated properties for other fiber volume fractions. Axial strength and modulus are strongly dependent on fiber volume fraction, v_f, and less so other properties. Let the ratio of the actual fiber volume fraction, v_{fa}, to the reference fiber volume fraction, $v_{fr} = 60\%$, be:

$$r_f = \frac{v_{fa}}{v_{fr}} \qquad (1.3\text{-}7)$$

Approximate properties for the material with fiber volume fraction, v_{fa}, are

- Longitudinal Modulus, $E_L = r_f$ times Table Value
- Transverse Modulus, E_T = Table Value
- Axial Shear Modulus, G_{LT} = Table Value
- Poisson's Ratio, ν_{LT} = Table Value
- Longitudinal Tensile and Compressive Strengths, F_L^{tu} and F_L^{cu} = r_f times Table Values
- Transverse Tensile and Compressive Strengths, F_T^{tu} and F_T^{cu} = Table Values
- In-Plane and Interlaminar Shear Strengths, F_L^{su} and F^{isu} = Table Values

Note that ultimate strains remain unchanged because both strengths and moduli are increased by the same percentage. Yield strength estimates follow the same rules as the corresponding ultimate strength values.

Some caution is in order when extrapolating data. The above rules generally provide reasonable preliminary design estimates over the range of fiber volume fractions from 50 to 70%. We note that the approximations assume that transverse, resin-dominated properties are constant over this range. This assumption is not correct, but at least for most preliminary designs, axial properties are of primary importance, and the variation in transverse properties can be neglected. If transverse properties are important for a particular application, then a more accurate estimate should be used. The analytical methods of Volume 2 present methods for obtaining these property estimates.

Particular care should be exercised in extrapolating strengths to higher values of fiber volume fraction. As illustrated in section 1.2, internal stress concentrations

Table 1.3-7. Representative properties of Kevlar 49 aramid/epoxy unidirectional composites: fiber volume fraction, $v_f = 0.60$.

	GPa	10^6 psi
Elastic Constants		
Longitudinal Modulus, E_L	76	11
Transverse Modulus, E_T	5.5	0.8
Axial Shear Modulus, G_{LT}	2.1	0.3
Poisson's Ratio, ν_{LT} (dimensionless)	0.34	
Strength Properties	MPa	10^3 psi
Longitudinal Tension, F_L^{tu}	1240	180
Longitudinal Compression, "Yield," F_L^{cy}	230	33
Longitudinal Compression, Ultimate, F_L^{cu}	280	40
Transverse Tension, F_T^{tu}	30	4.3
Transverse Compression, F_T^{cu}	140	20
In-Plane Shear, F_L^{su}	60	9
Interlaminar Shear, F^{isu}	60	9
Ultimate Strains (Percent)		
Longitudinal Tension, ϵ_L^{tu}	1.6	
Longitudinal Compression, "Yield," ϵ_L^{cy}	0.3	
Longitudinal Compression, Ultimate, ϵ_L^{cu}	>2.0	
Transverse Tension, ϵ_T^{tu}	0.5	
Transverse Compression, ϵ_T^{cu}	2.5	
In-Plane Shear, γ_{LT}^u	1–6	
Density, kg/m³ (lb/in³)	1.38×10^3 (0.050)	

for transverse extensional and shear stress increase with v_f. Further, for many fabrication processes, void content tends to increase as resin content decreases. Both of these factors can reduce transverse tensile and in-plane shear strengths. This strength reduction does not appear to be a serious problem for well-made composites, such as those made from prepreg under carefully controlled conditions.

The system dependence on extrapolation rules also applies to axial strength properties. Generally, they follow a simple rule of mixtures, as indicated in the extrapolation rules presented above. However, for some materials, strength properties drop off at higher fiber volume fractions, apparently because of poor fiber wetout, voids, etc. This strength reduction is particularly true for some composites made by processes such as filament winding. Figure 1.3-16 shows how fiber stress at failure varied with fiber volume fraction for E glass/epoxy strands in one particular study [40].

Table 1.3-8. Representative properties of high-strength graphite/epoxy unidirectional composites: fiber volume fraction, $v_f = 0.60$.

	GPa	10^6 psi
Elastic Constants		
Longitudinal Modulus, E_L	145	21
Transverse Modulus, E_T	10	1.5
Axial Shear Modulus, G_{LT}	4.8	0.7
Poisson's Ratio, ν_{LT} (dimensionless)	0.25	
Strength Properties	**MPa**	**10^3 psi**
Longitudinal Tension, F_L^{tu}	1240	180
Longitudinal Compression, F_L^{cu}	1240	180
Transverse Tension, F_T^{tu}	41	6
Transverse Compression, F_T^{cu}	170	25
In-Plane Shear, F_L^{su}	80	12
Interlaminar Shear, F^{isu}	80	12
Ultimate Strains (Percent)		
Longitudinal Tension, ϵ_L^{tu}	0.9	
Longitudinal Compression, ϵ_L^{cu}	0.9	
Transverse Tension, ϵ_T^{tu}	0.4	
Transverse Compression, ϵ_T^{cu}	1.6	
In-Plane Shear, γ_{LT}^u	1–6	
Density, kg/m³ (lb/in³)	1.58×10^3 (0.057)	

Table 1.3-10. Representative properties of ultrahigh-modulus graphite/epoxy unidirectional composites: fiber volume fraction, $v_f = 0.60$.

	GPa	10^6 psi
Elastic Constants		
Longitudinal Modulus, E_L	290	42
Transverse Modulus, E_T	6.2	0.9
Axial Shear Modulus, G_{LT}	4.8	0.7
Poisson's Ratio, ν_{LT} (dimensionless)	0.25	
Strength Properties	**MPa**	**10^3 psi**
Longitudinal Tension, F_L^{tu}	620	90
Longitudinal Compression, F_L^{cu}	620	90
Transverse Tension, F_T^{tu}	21	3
Transverse Compression, F_T^{cu}	170	25
In-Plane Shear, F_L^{su}	60	9
Interlaminar Shear, F^{isu}	60	9
Ultimate Strains (Percent)		
Longitudinal Tension, ϵ_L^{tu}	0.2	
Longitudinal Compression, ϵ_L^{cu}	0.2	
Transverse Tension, ϵ_T^{tu}	0.3	
Transverse Compression, ϵ_T^{cu}	2.8	
In-Plane Shear, γ_{LT}^u	0.6–4	
Density, kg/m³ (lb/in³)	170×10^3 (0.061)	

Table 1.3-9. Representative properties of high-modulus graphite/epoxy unidirectional composites: fiber volume fraction, $v_f = 0.60$.

	GPa	10^6 psi
Elastic Constants		
Longitudinal Modulus, E_L	220	32
Transverse Modulus, E_T	6.9	1.0
Axial Shear Modulus, G_{LT}	4.8	0.7
Poisson's Ratio, ν_{LT} (dimensionless)	0.25	
Strength Properties	**MPa**	**10^3 psi**
Longitudinal Tension, F_L^{tu}	760	110
Longitudinal Compression, F_L^{cu}	690	100
Transverse Tension, F_T^{tu}	28	4
Transverse Compression, F_T^{cu}	170	25
In-Plane Shear, F_L^{su}	70	10
Interlaminar Shear, F^{isu}	70	10
Ultimate Strains (Percent)		
Longitudinal Tension, ϵ_L^{tu}	0.3	
Longitudinal Compression, ϵ_L^{cu}	0.3	
Transverse Tension, ϵ_T^{tu}	0.4	
Transverse Compression, ϵ_T^{cu}	2.8	
In-Plane Shear, γ_{LT}^u	1–6	
Density, kg/m³ (lb/in³)	1.64×10^3 (0.059)	

Table 1.3-11. Representative properties of boron/epoxy unidirectional composites: fiber volume fraction, $v_f = 0.60$.

	GPa	10^6 psi
Elastic Constants		
Longitudinal Modulus, E_L	210	30
Transverse Modulus, E_T	19	2.7
Axial Shear Modulus, G_{LT}	4.8	0.7
Poisson's Ratio, ν_{LT} (dimensionless)	0.25	
Strength Properties	**MPa**	**10^3 psi**
Longitudinal Tension, F_L^{tu}	1240	180
Longitudinal Compression, F_L^{cu}	3310	480
Transverse Tension, F_T^{tu}	70	10
Transverse Compression, F_T^{cu}	280	40
In-Plane Shear, F_L^{su}	90	13
Interlaminar Shear, F^{isu}	90	13
Ultimate Strains (Percent)		
Longitudinal Tension, ϵ_L^{tu}	0.6	
Longitudinal Compression, ϵ_L^{cu}	1.6	
Transverse Tension, ϵ_T^{tu}	0.4	
Transverse Compression, ϵ_T^{cu}	1.5	
In-Plane Shear, γ_{LT}^u	1–6	
Density, kg/m³ (lb/in³)	2.0×10^3 (0.073)	

Table 1.3-12. *Representative properties of alumina/epoxy unidirectional composites: fiber volume fraction, $v_f = 0.60$.*

	GPa	10^6 psi
Elastic Constants		
Longitudinal Modulus, E_L	230	33
Transverse Modulus, E_T	21	3
Axial Shear Modulus, G_{LT}	7	1
Poisson's Ratio, ν_{LT} (dimensionless)	0.28	
Strength Properties	**MPa**	**10^3 psi**
Longitudinal Tension, F_L^{tu}	520	75
Longitudinal Compression, F_L^{cu}	2340	340
Transverse Tension, F_T^{tu}	55	8
Transverse Compression, F_T^{cu}	140	20
In-Plane Shear, F_L^{su}	41	6
Interlaminar Shear, F^{isu}	41	6
Ultimate Strains (Percent)		
Longitudinal Tension, ϵ_L^{tu}	0.2	
Longitudinal Compression, ϵ_L^{cu}	1.0	
Transverse Tension, ϵ_T^{tu}	0.4	
Transverse Compression, ϵ_T^{cu}	0.6	
In-Plane Shear, γ_{LT}^u	1–6	
Density, kg/m^3 (lb/in³)	2.8×10^3 (0.101)	

Since the data presented in this section are representative values, the question arises concerning what to use for preliminary design allowables. The resolution of the question will depend strongly on the manufacturing method used and the degree of quality control employed throughout the process, including testing of incoming materials and monitoring the condition of those in stock. As a reasonable approximation, we suggest the following reductions to obtain estimated strength design allowables:

"B" Value — 25% of average
"A" Value — 50% of average

Unidirectional composite properties are presented as follows:

- Table 1.3-5, E glass/epoxy
- Table 1.3-6, S glass/epoxy
- Table 1.3-7, Kevlar 49 aramid/epoxy
- Table 1.3-8, High-strength graphite/epoxy
- Table 1.3-9, High-modulus graphite/epoxy
- Table 1.3-10, Ultrahigh-modulus graphite/epoxy
- Table 1.3-11, Boron/epoxy
- Table 1.3-12, Alumina/epoxy

1.3.4 References

1. HASHIN, Z. "Analysis of Composite Materials—A Survey," *J. Applied Mechanics*, 50:481 (1983).

2. ROSEN, B. W. and Z. Hashin. "Analysis of Material Properties," *Engineered Materials Handbook*, ASM International, 1:185 (1987).

3. WHITNEY, J. M. "Elasticity Analysis of Orthotropic Beams Under Concentrated Loads," *Composites Science and Technology*, 22:167 (1985).

4. ZWEBEN, C., W. S. Smith and M. W. Wardle. "Test Methods for Fiber Tensile Strength, Composite Flexural Modulus and Properties of Fabric-Reinforced Laminates," *Composite Materials: Testing and Design*, ASTM STP 674, American Society for Testing and Materials, Philadelphia, p. 228 (1979).

5. YEOW, Y. T. and H. F. Brinson. "A Comparison of Simple Shear Characterization Methods for Composite Laminates," *Composites*, 9(1):49 (1978).

6. KNOFF, W. F. "A Modified Weakest-Link Model for Describing Strength Variability of Kevlar Aramid Fibers," *J. Materials Science*, 22:1024 (1987).

7. "Design Properties of Marine Grade Fiberglass Laminates," Owens-Corning Fiberglas Corporation, Publication No. 5-B0-5905-A (March 1973).

8. CARLSSON, L. A. and R. B. Pipes. *Experimental Characterization of Advanced Composite Materials*. Prentice-Hall, Inc., Englewood Cliffs, NJ (1987).

9. HALPIN, J. C., K. L. Jerina and T. A. Johnson. "Characterization of Composites for the Purpose of Reliability Evaluation," *Analysis of Test Methods for High Modulus Fibers and Composites*, ASTM STP 521, American Society for Testing and Materials, Phila., p. 5 (1973).

10. JONES, B. H. "Probabilistic Design and Reliability," *Composite Materials*, C. C. Chamis, ed., Academic Press, New York, 8:33 (1974).

11. YANG, J. N. "Reliability Prediction for Composites Under Periodic Proof Tests in Service," *Composite Materials: Testing and Design (Fourth Conference)*, ASTM STP 617, American Society for Testing and Materials, Philadelphia, p. 272 (1977).

12. MAXWELL, R. E., R. H. Toland and C. W. Johnson. "Probabilistic Design of Composite Structures," *Composite Reliability*, ASTM STP 580, American Society for Testing and Materials, Philadelphia, p. 35 (1975).

13. LENOE, E. M. and D. Neal. "Structural Integrity Assessment of Filament-Wound Composites," *Composite Reliability*, ASTM STP 580, American Society for Testing and Materials, Philadelphia, p. 54 (1975).

14. YANG, J. N. "Reliability Prediction and Cost Optimization for Composites Including Periodic Proof Tests in Service," AFML-TR-76-224, U.S. Air Force Materials Laboratory (November 1976).

15. *Military Standardization Handbook*, "Metallic Materials and Elements for Aerospace Vehicle Structures," MIL-HDBK-5B, Department of Defense (September 1971).

16. LEHMAN, G. M., et al. *Advanced Composite Rudders for DC-10 Aircraft—Design Manufacturing and Ground Tests*, NASA CR-145068, National Aeronautics and Space Administration (1976).

17. KIES, J. A. "The Strength of Glass Fibers and the Failure of Filament Wound Pressure Vessels," NRL Report 6034, U.S. Naval Research Laboratory, Washington, D.C. (1963).

18. SABNIS, G. M. and S. Aroni. "Size Effect in Material Systems—The State of the Art," Conference on Structures, Solid Mechanics and Engineering Design in Civil Engineering Structures, Southampton (1969).

19. RIEDINGER, L. A., M. H. Kural and G. W. Reed, Jr. "Evaluation of the Potential Structure Performance of Composites," *Mechanics of Composite Materials, Proceedings of the Fifth Symposium on Naval Structural Mechanics*, F. W. Wendt, H. Liebowitz and N. Perrone, eds., Pergamon Press, Oxford, p. 481 (1970).

20. HITCHON, J. W. and D. C. Philips. "The Effect of Specimen Size on the Strength of CFRP," *Composites*, 9(2):119 (1978).

21. HARTER, H. L. "A Bibliography of Extreme-Value Theory," *International Statistical Review*, 46:279 (1978).

22. ZWEBEN, C. "Tensile Failure of Fiber Composites," *AIAA Journal*, 6(12):2325 (1968).

23. ZWEBEN, C. and B. W. Rosen. "A Statistical Theory of Material Strength with Application to Composite Materials," *Journal of the Mechanics and Physics of Solids*, 18:189 (1970).

24. ZWEBEN, C. "A Bounding Approach to the Strength of Composite Materials," *Engineering Fracture Mechanics*, 4:1 (1972).

25. HARLOW, D. G. and S. L. Phoenix. "The Chain-of-Bundles Probability Model for the Strength of Fibrous Materials I: Analysis and Conjectures," *Journal of Composite Materials*, 12:195 (1978).

26. HARLOW, D. G. and S. L. Phoenix. "The Chain-of-Bundles Probability Model for the Strength of Fibrous Materials II: A Numerical Study of Convergence," *Journal of Composite Materials*, 12:314 (1978).

27. HARLOW, D. G. "Properties of the Strength Distribution for Composite Materials," *Composite Materials: Testing and Design (Fifth Conference)*, ASTM STP 674, American Society for Testing and Materials, Philadelphia, p. 484 (1979).

28. CROWTHER, M. F. and M. S. Starkley. "Use of Weibull Sta-
tistics to Quantify Specimen Size Effects in Fatigue of GRP," *Composites Science and Technology*, 31:87 (1988).

29. BULLOCK, R. E. "Strength Ratios of Composite Materials in Flexure and in Tension," *Journal of Composite Materials*, 8:200 (1974).

30. WEIBULL, W. "A Statistical Theory of the Strength of Materials," *Ing. Vetenskaps. Akad. Handl.* (Royal Swedish Institute of Engineering Research Proceedings), NR 151 (1939).

31. WEIBULL, W. "A Statistical Distribution Function of Wide Applicability," *J. of Applied Mechanics*, 18:293 (1951).

32. JOHNSTON, N. J. and P. M. Hergenrother. "High Performance Thermoplastics: A Review of Neat Resin and Composite Properties," NASA TM 89104 (February 1987).

33. CARLILE, D. R., D. C. Leach, D. R. Moore and N. Zahlan. "Mechanical Properties of the Carbon Fiber/PEEK Composite APC-2/AS4 for Structural Applications," presented at ASTM Symposium on Advances in Thermoplastic Matrix Composite Materials, 19–20 October, 1987, Bal Harbour, Florida, submitted for inclusion in ASTM STP.

34. ZWEBEN, C. "The Flexural Strength of Aramid Fiber Composites," *Journal of Composite Materials*, 12:422 (October 1978).

35. LUO, S. Y. and T. W. Chou. "Finite Deformation and Nonlinear Elastic Behavior of Flexible Composites," *J. Applied Mechanics*, 55:149 (1988).

36. JONES, R. M. "Apparent Flexural Modulus and Strength of Multimodulus Materials," *J. Composite Materials*, 10:342 (1976).

37. GREENWOOD, J. H. and P. G. Rose. "Compressive Behavior of Kevlar 49 Fibres and Composites," *Journal of Materials Science*, 9:1809 (1974).

38. KULKARNI, S. V., J. S. Rice and B. W. Rosen. "An Investigation of the Compressive Strength of Kevlar 49/Epoxy Composites," *Composites*, 6:217 (September 1975).

39. SLEPETZ, J. M., T. F. Zagaeski and R. F. Novello. "In-Plane Shear Test for Composite Materials," AMMRC TR 78-30, U.S. Army Materials and Mechanics Research Center, Watertown, Massachusetts (July 1978).

40. STONE, R. G., T. T. Chiao, J. A. Rinde, L. S. Penn, L. L. Clements and E. Wu. "Fiber Composite Program for Flywheel Applications," Third Quarterly Progress Report, Report No. UCRL-50033-76-1, Lawrence Livermore Laboratory (May 1976).

SECTION 1.4

Fatigue of Composites

When subjected to cycle loading, composites, like metals, in general exhibit a fatigue sensitivity which results in degradation of performance. Whereas in metals fatigue failure is understood to be primarily a problem of crack initiation and growth, composite fatigue damage mechanisms are more complicated and less thoroughly understood.

There are two distinct classes of fiber reinforced composite materials, continuous fiber composites (CFC) and discontinuous fiber composites (DFC). As one might expect, the fatigue failure mechanisms associated with these two systems are different.

Fatigue failure of continuous fiber composite laminates consists of the initiation of ply cracks, crack multiplication and delamination, and the fracture of plies with fibers in the loading direction. The ply cracks are across the thickness of the lumped plies with the same fiber orientation and are parallel to the fibers. The nucleation of these cracks is instantaneous if fatigue stress is above the first ply failure stress. Otherwise, it will depend on the fatigue behavior of each constituent ply as well as on the state of ply stress. In the practical range of applied stresses, ply cracks are initiated rather early in fatigue life of laminates, and the crack multiplication takes up most of the fatigue life. The ply cracks also lead to delamination especially at free edges which are subjected to high interlaminar stresses. The ply failures and delamination manifest themselves in the change of mechanical properties such as modulus and strength. These subcritical failures no doubt depend on the type of laminate, loading condition, geometrical discontinuity, and environment.

The differences in the fatigue failure mechanism for discontinuous fiber composites result primarily from the modified microstructural characteristics. Crack initiation can occur at fiber ends as a result of stress concentrations developed there or at the fiber-matrix interface along fibers oriented perpendicular to the load direction. Once started, a crack usually propagates through the matrix until arrested by a fiber impeding the damage progression. In composites with relatively long fiber reinforcements, cracks can bridge fibers, leading to multiple cracking. As with continuous fiber composites the fatigue failure mechanism involves crack initiation and multiplication (in most cases) but the cracks are shorter, shallower and more randomly oriented and dispersed. Unlike the continuous fiber system which exhibits intense growth of damage early approaching asymptotic growth near failure, DFC exhibit slower crack growth early in the fatigue life with rapid growth near failure.

This section presents a mechanism-oriented, rather than material-oriented, discussion of fatigue behavior of composite laminates and discontinuous fiber composites. Macroscopic indications of fatigue damage are explained in terms of ply failures, delamination and macroscopic crack formation.

Fatigue life of composites is discussed in terms of S-N curves and residual strength. Several types of fatigue life models are described including models based on laminate strength analogies and statistical models. The statistical models employ the failure potential and strength degradation approaches to life prediction.

Nomenclature

a_T	time-temperature reduction factor
B/Al	boron/aluminum
Be/Al	beryllium/aluminum
B/Ep	boron/epoxy
BSiC/Ti	borsic/titanium
f	test frequency
$f(\ )$	failure function
F_i, F_{ij}	strength tensor components
FPF	first ply failure
Gl/Ep	glass/epoxy
Gr/Ep	graphite/epoxy

h	laminate thickness
Kv 49/Ep	Kevlar 49/epoxy
N	number of cycles to failure
n	number of cycles endured
N_x	laminate stress resultant in the 0-degree direction
N_{FPF}	laminate stress resultant at the first ply failure
R	stress ratio (= minimum fatigue stress/maximum fatigue stress)
$R(t)$	probability of surviving time t
S	applied rupture stress
S_{max}	maximum fatigue stress
$S_i(N)$	uniaxial fatigue strengths at N cycles
T	temperature, K
T_R	room temperature (= 296 K)
t	time, h
t_0	characteristic lifetime
UTS	(average) ultimate tensile strength
X_i	uniaxial static strengths
α	shape parameter for life distribution
ΔT_e	equilibrium temperature increase
σ_i	applied stress components
sub L	longitudinal
sub S	shear (longitudinal)
sub T	transverse
$[0/\pm 45/90]_s$	laminate stacking sequence
$(0/\pm 45/90)$	fiber orientations in a laminate
fatigue ratio	fatigue strength (mostly at 10^6 cycles) divided by UTS

1.4.1 Introduction

Fatigue failure in metals is known to consist of crack initiation and growth [1]. The crack initiation process starts with a period of work hardening. Submicrocracks then are formed at the ends of slip bands. These submicrocracks subsequently grow and coalesce to form a crack of detectable size, completing the crack initiation process. Upon further fatigue loading, the crack grows until fracture. The periods of work hardening and macroscopic crack growth are relatively short and most of fatigue life is spent in the formation and growth of submicrocracks [2].

Fatigue of composite laminates, on the other hand, is characterized by the initiation and, mostly, multiplication, not growth, of cracks. Crack initiation in compos-ite laminates coincides with the first ply failure, i.e., the first cracking of the weakest ply. After the first ply failure more cracks are formed in the same weakest ply. As fatigue proceeds further, cracks begin to appear in the next weakest plies. Some cracks in different plies are joined by delamination. Final fracture then follows when fibers in the loading direction break [3–5]. Thus, whereas fatigue failure of metals is a result of the initiation and growth of a single dominant crack, fatigue failure of composite laminates is preceded by multiple cracking of plies and delamination.

As mentioned earlier, the crack initiation process in metals takes much longer than the crack growth. However, the crack initiation in composite laminates takes place very early in fatigue life, and the crack multiplication and delamination, instead, take up most of fatigue life. Thus, cracks in composite laminates certainly have quite different implications from those in metals with regard to their criticality.

Whereas the crack growth in metals accelerates with fatigue, the crack multiplication in composite laminates decelerates. The stress intensity at crack tip in metals increases as the crack grows and the final fracture is the result of this accelerated crack growth. However, the ply stresses which are responsible for crack multiplication in composite laminates decrease as more cracks are formed. Also, cracks in weak plies do not seem to critically affect the load-bearing capability of strong plies, perhaps because of delamination (for the case of tensile loading).

The yield point for metals is macroscopically analogous to the first ply failure point for composite laminates. Microscopically, however, the yield point is associated with the incipience of massive dislocation motion while the first ply failure point represents the first cracking of the weakest plies. Yet, they both mark a transition from elastic behavior to inelastic behavior.

The similarities are welcome because we can borrow from the well-developed metals technology. The differences point up a need to develop a design methodology suitable for composites.

As is evident from several review papers [6–11], there are too many materials and structural variables to fully discuss details of fatigue behavior of various composites. Also, being a relatively new technology, fatigue of composites is not without many contradictions and variations which at present cannot be satisfactorily explained.

Therefore, in this section we provide a mechanism-oriented, rather than material-oriented, discussion of fatigue behavior of composites. To facilitate a clear understanding, common features are emphasized more than exceptions. Since unidirectional composites make up composite laminates, their behavior is discussed first under longitudinal and then under off-axis loadings. Also covered is the stress rupture of unidirectional composites. The discussion of laminate behavior includes failure processes, stress-life (*S-N*) relations, effect of compression, change of modulus, change of strength, temperature increase, frequency effect, fatigue notch sensitivity, effect of stacking sequence, and environmental effects. Metal-matrix composites are included only in the discussion of unidirectional composites. No discussion is given of their laminate behavior.

1.4.2 Unidirectional Continuous Fiber Composites

Longitudinal Tension Fatigue

The fibers used in modern composites are stronger and stiffer than the matrix materials; however, these fibers show much lower strain capability. As a result, failure of composites is initiated by fracture of fibers because both fibers and matrix are subjected to the same axial strain. The average stress-strain curves of the composite and its constituents are schematically shown in Figure 1.4-1. In the figure the effect of a brittle interface such as the reaction zone in BSiC/Ti [12,13] has been included in the fiber strength.

Although the average behavior can be understood from the schematic diagrams of Figure 1.4-1, the actual failure processes are much more complicated. Because of the variability in strength and imperfect alignment, some fibers will undoubtedly fail earlier than others. Furthermore, there may be fibers already broken before or during fabrication. What happens after a fiber breaks depends on the properties of the fiber, matrix, and interface. The pertinent mechanical properties include strength, stiffness, and toughness.

Four possible modes of subsequent fracture are shown in Figure 1.4-2, where phases I and II are fiber and matrix, respectively. In the first mode (a) the matrix is tough and the crack cannot grow into the matrix. Therefore, the fiber fractures at another loca-

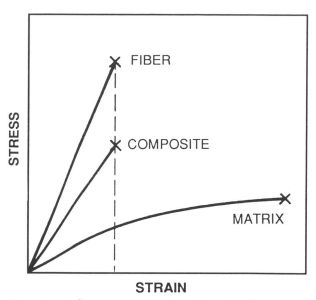

FIGURE 1.4-1. Schematic stress-strain relations for fiber, matrix and composite, respectively.

tion as the load is increased. The second mode (b) results when the interface is weak. The third mode (c) occurs when the matrix has low toughness and is sensitive to notches. Finally, the last mode (d) can happen if the matrix and interface are strong, stiff and tough [14,15]. In this case the stress concentration is large enough to cause fracture of the neighboring fiber without cracking of the matrix.

If the second mode (b) is the dominant failure mode, the composite will behave like a dry fiber bundle. On the other hand, the third mode (c) is reminiscent of crack propagation in brittle materials. Neither of these two is desirable because the average fiber strength is not fully utilized: in mode (b), the bundle strength is lower than the average fiber strength, and in mode (c), the composite strength is controlled by weak fibers. An optimum strength is obtained in failure mode (a) [16,17].

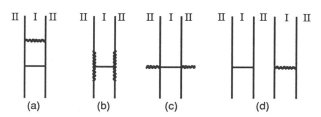

FIGURE 1.4-2. Crack growth modes in unidirectional composites.

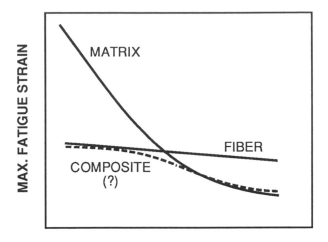

FIGURE 1.4-3. Schematic *S-N* relations for fiber, matrix and composite, respectively.

Similar types of crack growth are possible when a crack in the matrix approaches fibers. Figure 1.4-2 is still applicable if we interpret phase I as the matrix phase. The only difference is that the first mode (a) is not possible in static tension since the matrix has higher failure strain than the fiber. In fatigue, however, mode (a) is possible, as will be discussed later. The last mode (d) should be interpreted as a crack growth in the matrix around a fiber.

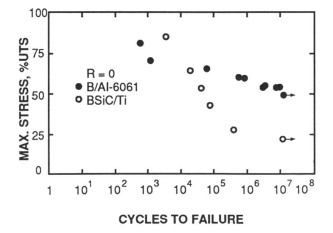

FIGURE 1.4-4. Longitudinal *S-N* data for B/Al-6061 and BSiC/Ti (UTS = 1698 MPa for B/Al and 1296 MPa for BSiC/Ti).

In addition to the mechanical properties of constituents just discussed, the subsequent crack growth also depends on the load level and loading rate at which the fiber break occurs. If the fiber breaks at a low stress due to defects or weakness, the crack is more likely to lead to interfacial debond than to grow into the matrix or to cause dynamic fracture of neighboring fibers [18–20]. Consequently, when a composite contains many weak fibers, a substantial number of fiber breaks are required before the composite fails. However, if all fibers are strong, a few fiber breaks seem to be enough to precipitate the composite failure [14].

Similarly, interfacial failure is more likely to occur at preexisting fiber ends [14,21]. This difference between preexisting fiber ends and new fiber breaks caused by loading makes it possible to increase the composite strength by breaking weak fibers before fabrication [20].

The subsequent crack growth into the matrix is also more likely at a higher loading rate than at a lower loading rate [18]. Thus, the composite strength may decrease with increasing loading rate [22].

The foregoing discussion of static failure modes is applicable to fatigue failure as well. However, fatigue failure requires one more parameter—the fatigue sensitivity of each constituent phase. Here the expression *fatigue sensitivity* is loosely defined as the degree of reduction of fatigue strength with increasing life.

Typical average stress-strain (*S-N*) relations of constituents are schematically shown in Figure 1.4-3. Maximum fatigue strain, rather than maximum fatigue stress, is used for easy comparison. In the low cycle region, the failure of the composite is triggered by failure of fibers as in static tension. However, in the high cycle region the matrix begins to fail before the fibers because of its higher fatigue sensitivity. Thus, the failure mode of Figure 1.4-2(a) with phase I as matrix is also possible. It is expected that, on the average, the fatigue limit strain of the composite would be about the same as that of the matrix material. However, the actual fatigue limit will depend on the presence of stress raisers such as early fiber breaks, fiber ends, brittle reaction zones, and voids.

Figure 1.4-4 shows an *S-N* relation of B/Al-6061, where maximum fatigue stress has been normalized with respect to the average ultimate tensile strength (UTS) [23]. Although boron fiber is known to be highly resistant to fatigue [24], the fatigue ratio at 10^7 cycles is only about 0.5.

Even in a tension-tension fatigue of $R = 0.1$, the aluminum matrix in the composite is subjected to a tension-compression fatigue. The resulting work hardening facilitates crack nucleation and growth in the matrix. The matrix shows evidence of work hardening and shear failure on fatigue fracture surfaces. However, only a ductile failure mode is observed on static fracture surfaces [25]. In fact, the shakedown of aluminum matrix was shown to lead to composite failure [26].

On a microscopic scale, fatigue cracks in the matrix can be initiated at free surfaces as in metals [9,27]. Also, weak fibers can fail not only in preloading at a stress as low as only 50 percent of the strength [14,27], but also in fatigue [9]. These fiber breaks, as well as preexisting fiber ends, lead to more fiber fractures through a combination of crack growth modes (c) and (d) in Figure 1.4-2 [9,14]. When there are enough fiber breaks (some of them connected by matrix cracks) within a zone so that their effects are additive, the composite fails [3].

If fibers are ductile, a crack growth of mode (d) is not likely to occur. Instead, a stable crack growth of modes (b) and (c) is possible, leaving fatigue striations as observed in Be/Al [28].

Crack growth of mode (c) can be retarded by using an aluminum with lower yield stress, thus leading to an increased fatigue limit. In Be/Al-1235, cracks frequently grow along the interface, whereas virtually no branching of fatigue cracks occurs in Be/Al-6061-MT6. The resulting fatigue ratio for Be/Al-1235 is much higher than for Be/Al-6061-MT6 [9].

Brittle coatings such as SiC and brittle reaction zones at the interface substantially reduce the composite strength [12]. However, the effect of a brittle interface on the fatigue limit can be relatively small if the interfacial bond is weak [9]. Otherwise, the fatigue limit is degraded also.

Figure 1.4-4 also shows an *S-N* relation of BSiC/Ti-6Al-4V. Here the fatigue ratio is much lower than that of B/Al-6061. Since titanium has higher stiffness and strength than aluminum, the load transfer from a broken fiber to an intact neighboring fiber is more effective and the transverse crack growth of mode (d) in Figure 1.4-2 is more likely. At the same time it may take some time from one fiber fracture to the next. The titanium matrix between the two broken fibers undergoes an accelerated work hardening and fails early [29]. The net result is a low fatigue ratio for the composite.

The fatigue of Gl/Ep composites is similar to that of B/Al composites in that the matrix itself undergoes a substantial damage in fatigue. The fatigue cracks in the matrix can be observed on the edges of the specimen [30,31]. Figure 1.4-5 compares the fatigue strengths of bulk epoxy and composites in terms of maximum fatigue strain. Even in the low-cycle region the matrix is expected to fail before the fibers. However, because of the weak interface (compared with B/Al) and lower stiffness of the epoxy, such cracks do not lead to immediate fracture of glass fibers. Consequently, the composite fails much later than the matrix. However, the difference disappears in the high-cycle region; the fatigue limit strain of the composite is close to that of the matrix.

In Gr/Ep and B/Ep composites the static failure strain is in the order of 1 percent which is close to the fatigue limit strain of the epoxy in Figure 1.4-5. Consequently, the resulting fatigue ratios are expected to be high. Figure 1.4-6 shows an *S-N* relation of high-modulus Gr/Ep which has an average failure strain of only 0.5 percent. The corresponding fatigue limit is only slightly below the lower tail end of the static strength distribution [8,32].

On the other hand, high-strength Gr/Ep with a larger average failure strain of 1 percent shows a slightly lower fatigue ratio, as expected, in Figure 1.4-7 [33]. The fatigue ratio of B/Ep shown in Figure 1.4-8 [34] falls between those of high-modulus Gr/Ep and high-strength Gr/Ep.

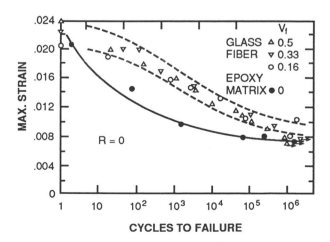

FIGURE 1.4-5. Longitudinal *S-N* data for Gl/Ep composites and epoxy matrix. Maximum fatigue strain is used on the ordinate.

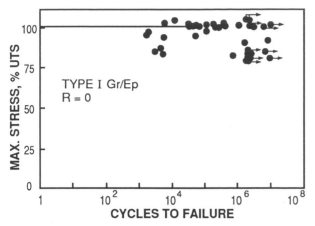

FIGURE 1.4-6. Longitudinal *S-N* data for high-modulus Gr/Ep (UTS = 872 MPa).

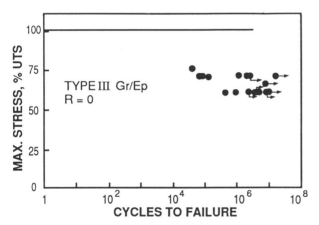

FIGURE 1.4-7. Longitudinal *S-N* data for high-strength Gr/Ep (UTS = 1732 MPa).

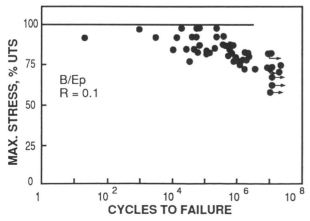

FIGURE 1.4-8. Longitudinal *S-N* data for B/Ep (UTS = 1331 MPa).

An *S-N* relation of Kevlar 49/epoxy is shown in Figure 1.4-9 [35]. This composite is seen to exhibit almost the same fatigue sensitivity as B/Ep or high-strength Gr/Ep, although the static failure strain is in the range of 2 percent. Two factors can be cited as responsible for such high fatigue resistance. One is that fracture of Kevlar fiber is not through a smooth surface normal to the fiber axis. It is accompanied by extensive fiber splitting [36–40]. Consequently, no sharp cracks are produced by fiber breaks and no transverse crack growth is permitted across fibers. The other factor is the weak interface, as evidenced by low transverse tensile and shear strengths [36,41]. Both of these two factors restrict transverse crack growth in the composite, and the composite is highly resistant to fatigue. In fact, the normalized *S-N* relation of the composite is very similar to that of Kevlar fiber itself [36,37].

In polymer-matrix composites, final failure is usually preceded by longitudinal splitting. In some cases the longitudinal cracks grow into the end tab regions, initiating a tab debonding [42,43]. However, a quantitative delineation of the effect of tab failure on life distribution is not easy because the life accompanied by tab failure is well within the scatter band of the "legitimate" data. Yet the data can be taken as a lower bound of a "real" *S-N* relation.

The actual state of stress within each phase of a composite is not uniaxial, even when the composite is under a longitudinal tension [44]. In particular, the radial stress at the interface can be tensile in the matrix-rich region. Also, the circumferential stress in the matrix at the interface of a Gr/Ep composite can be as high as 9 percent of the applied stress. Compounding these stresses are the curing stresses resulting from thermal expansion mismatch between fibers and matrix [44–46]. The combined stresses, in addition to fiber breaks, are conducive to longitudinal cracking [47].

Macroscopic fracture characteristics of composites in fatigue are not much different from those in static tension. The fracture surface of BSiC/Ti is fairly smooth and normal to the loading [29]. However, frequent jogs along the fibers are observed in B/Al [26]. The fracture surface becomes more irregular in Gr/Ep [10]. Both Gl/Ep and Kv/Ep composites show brush-like failures [36,43]. Certainly, fiber surface treatment reduces longitudinal cracking [8,32,47]. However, it may not result in improvements in static and fatigue strengths if it promotes too much of the crack growth mode (c) in Figure 1.4-2 [8,48].

With the exception of BSiC/Ti [29], no macroscopic growth of a transverse crack is observed, and failure is rather sudden. Also, neither modulus nor strength shows much reduction unless the effective cross-sectional area is reduced by longitudinal cracks [9,42, 49–51]. Thus, the final failure process is a dynamic fracture. The brush-like failure mode is a combined result of elastic energy being released suddenly, a weak interface, and a brittle matrix. The maximum strain energy densities of glass and Kevlar fibers before failure are more than twice as high as those of graphite and boron fibers.

Longitudinal Stress Rupture

Creep of unidirectional composites subjected to longitudinal tension mostly consists of two regions: a first-stage creep region of decreasing rate followed by a second-stage steady state creep region [25,29]. In the first-stage region, load is transferred from matrix to fibers as a result of the creep of the matrix. The steady state region is controlled mostly by the creep of fibers. Brittle fibers such as used in modern composites under discussion are not very susceptible to creep. Therefore, the third-state creep of increasing rate is not usually observed in composites.

Although the creep rupture strength of composites is mainly controlled by the fibers [25,52], the matrix and the interface can make a difference [53]. Because of the nonuniformity in fiber strength, some fibers will fail earlier than others. If fractures of the two neighboring fibers are far apart in the fiber direction, the stress concentration at one fiber break is not influenced by the other. However, when the two fiber breaks are connected by crack propagation along the fibers or by stress relaxation in the matrix, the resulting stress concentration will be close to what it would be if the two fiber breaks were on the same plane and connected [54]. Final fracture may then result because of higher stress concentration.

The longitudinal crack growth and stress relaxation certainly depend on the matrix and interface properties, and hence stress-rupture of composites will also be influenced by the matrix and interface.

Figure 1.4-10 shows stress-rupture data of B/Al-6061 and BSiC/Ti composites at several different temperatures [13,25,29]. There is very little difference among the rupture curves except at the highest temperature of 538°C.

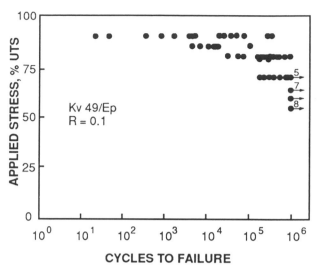

FIGURE 1.4-9. Longitudinal *S-N* data for Kv 49/Ep (UTS = 1586 MPa) (numbers of run-outs are shown at 10^6 cycles).

Figure 1.4-11 shows relations between applied stress and lifetime at 10 percent failure probability for several composite strands. Gr/Ep strand is seen to be least susceptible to delayed failure while Gl/Ep shows the most reduction in strength with time. The tests were carried out in room environment [55].

As in fatigue, the stress-rupture lifetime exhibits large scatter, necessitating a statistical analysis of data [56]. One of the widely used functions is the two-parameter Weibull distribution of the form:

$$R(t) = \exp\left[-\left(\frac{t}{t_0} \right)^{\alpha} \right] \qquad (1.4\text{-}1)$$

Here α and t_0 are called the shape parameter and characteristic lifetime, respectively. t is lifetime and $R(t)$ represents the probability of surviving time t. A method of determining α and t_0 from experimental data is described in the Appendix.

For Kv 49/Ep and S-Gl/Ep strands in Table 1.4-1, α and t_0 are shown in Figures 1.4-12 and 1.4-13, respectively [57]. At each stress level more than 45 strands were tested. The open symbols represent censored data and the closed symbols are for complete data. For the Kv 49/Ep composites, the underlying failure process seems to change as the applied stress crosses 80% of the ultimate tensile strength (UTS). Above 80% UTS the shape parameter is smaller than unity and the stress-lifetime relation is flatter. Below 80% UTS the

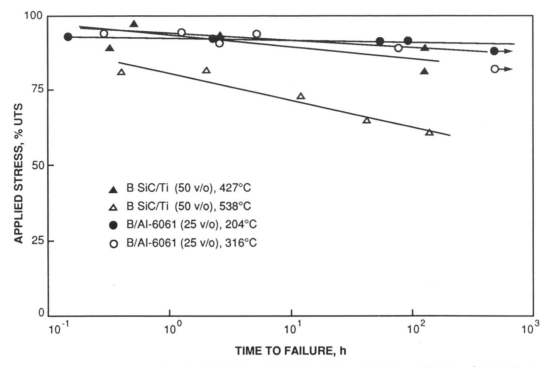

FIGURE 1.4-10. Stress rupture data for B/Al-6061 and BSiC/Ti (UTS = 592 MPa at 204°C and 548 MPa at 316°C for B/Al; UTS = 993 MPa at 427°C and 855 MPa at 538°C for BSiC/Ti).

shape parameter becomes larger than unity while the stress-lifetime relation assumes larger slope. Thus, the stress rupture above 80% UTS is controlled by initial defects, whereas these defects play only a secondary role at lower stresses. At higher stresses, a fiber break has more tendency to propagate normal to the neighboring fibers [18], and hence the probability of failure is higher. Once a strand survives this initial period, however, the fiber break will lead to interfacial failure and the probability of failure decreases. At lower stresses such change of failure mode does not take place and failure of the composite is a result of the wear-out of Kevlar fibers.

The downturn in Figure 1.4-13 observed of Kv 49/Ep-B, the data of which are also shown in Figure 1.4-11, is probably a manifestation of a delayed degradation resulting from the long-term exposure of stress-rupture specimens to fluorescent lights in the test laboratory [57]. Kevlar 49 fibers absorb light in the ultraviolet region, and such absorption can cause degradation of strength [58–60]. Unstressed strands suffered about 15 percent reduction in strength after 2¾ years of storage in the room where stress rupture tests were performed [55]. However, other Kv 49/Ep strands did not show any degradation after up to 3½ years of storage when shielded from light in a well-controlled room [61]. Stress-rupture testing of Kv 49/Ep-A was carried out in a room which was kept dark most of the time. Below 80% UTS this composite does not show any sign of downturn in its stress-logarithmic lifetime relation.

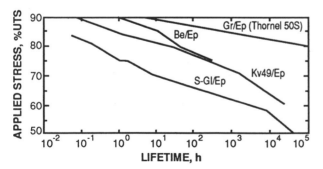

FIGURE 1.4-11. Stress-rupture curves of several composite strands at 10% failure probability [UTS – 3.87 GPa for S glass, 3.48 GPa for Kevlar 49, 1.97 GPa for Thornel 50S, and 1.16 GPa for Beryllium (Thornel is a registered trademark of Narmco Materials, Inc.)].

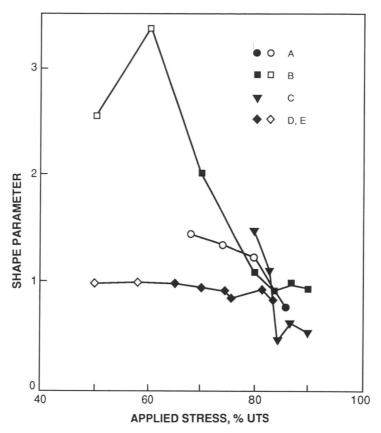

FIGURE 1.4-12. Effect of applied stress on shape parameter for composites listed in Table 1.4-1.

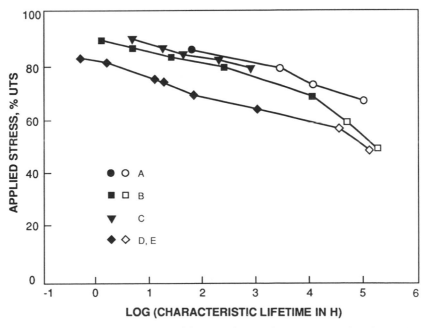

FIGURE 1.4-13. Stress-characteristic lifetime relations for composites listed in Table 1.4-1.

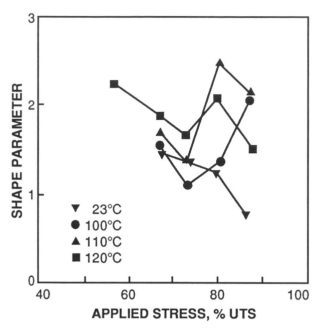

FIGURE 1.4-14. Shape parameters at elevated temperatures.

The stress-rupture characteristics of Kv 49/Ep-A at room temperature (RT) are compared with those of Kv 49/Ep-F at three elevated temperatures (ET) in Figures 1.4-14 and 1.4-15. Unlike the RT data, the shape parameters at ET do not depend on applied stress. Also, the RT shape parameters at 80% UTS and below are comparable to the ET shape parameters, indicating a similarity between the RT failure process at lower stresses and the ET failure process. Thus, all the data can be pooled together with the resulting joint shape parameter of 1.643 [57].

The stress-log t_0 relations at different temperatures are parallel to one another (Figure 1.4-15). Thus, the ET data can be combined with the RT data by horizontal shifting. The appropriate equation is:

$$\frac{S}{\text{UTS}} = -8.01 \times 10^{-2} \log (t_0/a_T) + 1.065 \quad (1.4\text{-}2)$$

where S is the applied stress [57]. The time-temperature reduction factor a_T depends on temperature T by:

$$-\log a_T = 4.83 \times 10^3 \left(\frac{1}{T_R} - \frac{1}{T} \right) \quad (1.4\text{-}3)$$

where T_R is the room temperature ($=296$ K). Equations (1.4-2) and (1.4-3) are shown with the experimental data in Figures 1.4-16 and 1.4-17, respectively. Thus, for the Kv 49/Ep composite the time-temperature reduction can be used to predict long-term lifetime from short-term data at an elevated temperature.

Off-Axis Fatigue

Off-axis fatigue behavior of unidirectional composites is as much controlled by the matrix and interface as off-axis strength is.

In metal-matrix composites reinforced with boron or Borsic fibers, composite failure is frequently initiated by splitting of fibers [13,45,49]. Increased transverse strength of larger diameter fibers can lead to a substantial improvement of static strength. However, in the high-cycle region, composite failure is commonly a result of fatigue-induced degradation of the matrix and interface. Therefore, the improvement in the fatigue limit that can result from fibers with higher transverse strength is less than that in the static strength [25,45]. Also, the use of heat-treated aluminum improves the static strength more than the fatigue limit [25].

Fatigue hardening of a metal matrix [62] leads to an increase in residual strength [23,49]. The fact that no decrease in residual strength occurs until close to final failure indicates that cracks are formed and grow fast in the last stage of fatigue life.

Table 1.4-1. Material systems in Figures 1.4-12 and 1.4-13.

Material System	Fiber	Epoxy Composition	Strength,[a] GPa
A	Kevlar 49	DER332/T403 (100/45)	3.24
B	PRD-49-III (Kevlar 49)	ERL2258/ZZL0820 (100/30)	3.48
C	Kevlar 49	XD7818/XD757.02/ XD7114/Tonox 60-40 (50/50/50/35.8)	3.43
D	S glass	DER332/T403 (100/36)	3.88
E	S glass	DER332/ERL4206/ EPI-Cure 855 (70/30/40)	3.94
F	Kevlar 49	ERL 2258/ZZL0820 (100/30)	3.36

[a]Fiber strength calculated from the average failure load of strands.

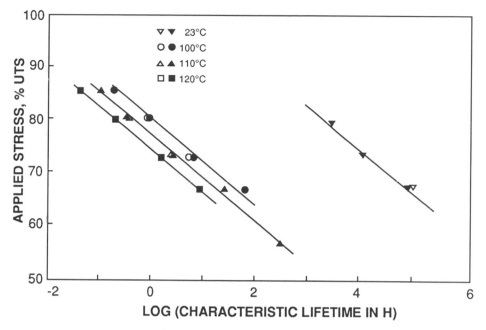

FIGURE 1.4-15. Characteristic lifetimes at elevated temperatures.

Figure 1.4-18 shows a transverse *S-N* relation for B/Ep [22,63]. Other polymer-matrix composites show similar behavior [64,65]. Typical transverse fatigue properties of composites are listed in Table 1.4-2. Note that the metal-matrix composites have higher static and fatigue strengths than do the polymer-matrix composites.

The fatigue sensitivity of (±45) laminates is similar to that in transverse tension fatigue [52,65,66,67]. These laminates are frequently used to determine the longitudinal shear properties. When the longitudinal-to-transverse modulus ratio is large, as in B/Ep and Gr/Ep, the transverse stress in the plies is much smaller than the longitudinal shear stress and may, therefore, be neglected. Otherwise, the transverse stress can be large enough to affect failure.

Figure 1.4-19 shows the longitudinal shear fatigue data for B/Ep obtained from a (±45) laminate [63]. The figure also includes the interlaminar shear fatigue data from reference [68]. Typical fatigue properties of (±45) laminates are listed in Table 1.4-3.

Longitudinal shear properties can also be obtained from 10-degree off-axis tension tests of unidirectional composites [69,70]. However, there is a major difference between the two methods, especially in fatigue of

polymer-matrix composites. Whereas (±45) laminates show gradual degradation in modulus and strength for the reasons to be discussed in the next subsection, 10-degree off-axis specimens may not [66,67,71]. For example, all off-axis fatigue failures of a Gr/Ep composite occurred without any sign of reduction in modulus or strength [71].

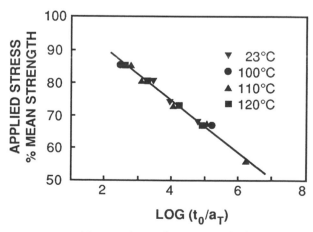

FIGURE 1.4-16. Master relation between applied stress and reduced characteristic lifetime.

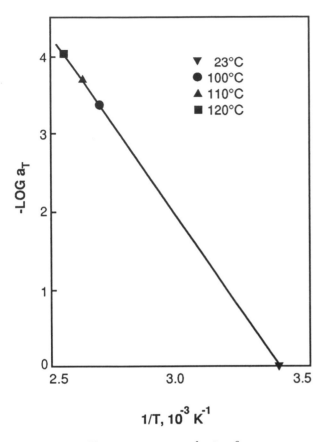

FIGURE 1.4-17. Time-temperature reduction factor a_T versus temperature.

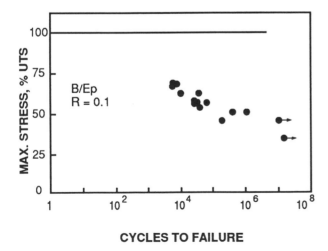

FIGURE 1.4-18. Transverse S-N data for B/Ep (UTS = 61 MPa).

Table 1.4-2. Transverse fatigue properties.

Material	Static Strength, MPa	Stress Ratio	Fatigue[a] Limit, MPa	Reference
B_4/Al-6061 (60 v/o)	108	0.4	89	[25]
B_4/Al-6061 T6M (60 v/o)	200	0.4	110	[25]
$B_{5.6}$/Al-6061 (60 v/o)	159	0.4	113	[25]
$B_{5.6}$/Al-6061 T6 (60 v/o)	306	0.4	156	[25]
BSiC/Ti-6Al-4V (38 v/o)	386	0.1	162	[66]
B/Ep	61	0.1	24	[22]
Gr/Ep	40	0.1	27[b]	[67]
E-Gl/Ep	29	0.1	18[b]	[64]
S-Gl/Ep	57	0.1	27	[65]

[a]At 10^7 cycles.
[b]At 10^6 cycles.

Off-axis fatigue strength can be predicted in the same way as off-axis static strength can. Various failure criteria for unidirectional composites have been reviewed in references [72,73]. In general, a failure criterion under a combined state of stress, σ_i ($i = L, T, S$, where L denotes "longitudinal," T denotes "transverse," and S denotes "shear"), is expressed as

$$f(\sigma_i; X_i) = 1 \qquad (1.4\text{-}4)$$

where X_i usually are the uniaxial strengths in the material symmetry axes. In fatigue, X_i are replaced by fatigue strengths $S_i(N)$ at a given number of cycles, so that

$$f[\sigma_i; S_i(N)] = 1 \qquad (1.4\text{-}5)$$

Although any of the available static failure criteria can be applied, the following four have been examined in the literature [64,65,72,74]:

$$\max\left\{ \frac{\sigma_L}{S_L}, \frac{\sigma_T}{S_T}, \frac{\sigma_s}{S_s} \right\} = 1 \qquad (1.4\text{-}6)$$

$$\max\left\{ \frac{\sigma_L}{S_L}, \left(\frac{\sigma_T^2}{S_T^2} + \frac{\sigma_S^2}{S_S^2} \right) \right\} = 1 \qquad (1.4\text{-}7)$$

$$\frac{\sigma_L^2}{S_L^2} - \frac{\sigma_L \sigma_T}{S_L^2} + \frac{\sigma_T^2}{S_T^2} + \frac{\sigma_S^2}{S_S^2} = 1 \qquad (1.4\text{-}8)$$

$$F_i \sigma_i + F_{ij} \sigma_i \sigma_j = 1 \quad (\text{summation over } i \text{ and } j) \qquad (1.4\text{-}9)$$

The first three were used for fatigue of Gl/Ep, and the last for stress rupture of Gl/Ep.

According to Equations (1.4-6) and (1.4-7), the off-axis fatigue-to-static strength ratio for matrix-controlled failure is independent of off-axis angle if $S_T/X_T = S_S/X_S$. Fortunately, most composites show only a weak dependence of off-axis fatigue limit on off-axis angle [25,64,71]. Thus, once uniaxial fatigue limits are known, off-axis fatigue strengths can easily be estimated.

1.4.3 Multidirectional Laminates

Failure Processes

A constituent ply in a laminate can fail in two different ways: parallel to fibers (called parallel failure) and normal to fibers (called normal failure). In the former, failure is mostly in the matrix and interface along the fibers and in the latter, fibers fail. In the absence of fibers in the loading direction, the parallel failure can lead to final failure of the composite. Otherwise, final failure of the composite coincides with the failure of fibers in the loading direction. Typical parallel ply failures observed on an edge are shown in Figure 1.4-20 for $[0/90]_s$, $[0/\pm45]_s$, and $[0/\pm45/90]_s$ laminates.

One group of laminates where the parallel ply failure is the dominant failure mode is ($\pm\theta$) angle-ply laminates subjected to a load in the 0-degree direction. The uniaxial tensile strengths of these laminates are compared with the off-axis strengths of unidirectional B/Ep laminae in Figure 1.4-21 [75]. Also shown are the predictions from the maximum stress criterion.

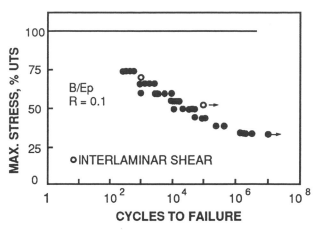

FIGURE 1.4-19. Shear *S-N* data for B/Ep (longitudinal shear strength = 67 MPa, interlaminar shear strength = 81 MPa).

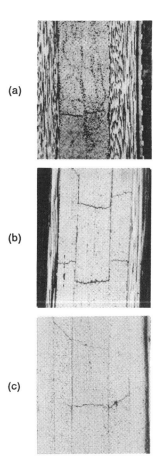

FIGURE 1.4-20. Ply failures observed on edge of Gr/Ep laminates: (a) $[0/90]_s$; (b) $[0/\pm45]_s$; (c) $[0/\pm45/90]_s$.

Table 1.4-3. Fatigue properties of (±45) laminates.

Material	Static Strength, MPa	Stress Ratio	Fatigue[a] Limit, MPa	Reference
B₄/Al-6061 (60 v/o)	160	0.4	114	[52]
B₄Al-6061 T6M (60 v/o)	206	0.4	150	[52]
B/Ep	133	0.1	40	[34]
Gr/Ep	196	0.1	78[b]	[66]
E-Gl/Ep	177	0.1	81[b]	[67]
S-Gl/Ep	186	0.1	71	[65]

[a]At 10^7 cycles.
[b]At 10^6 cycles.

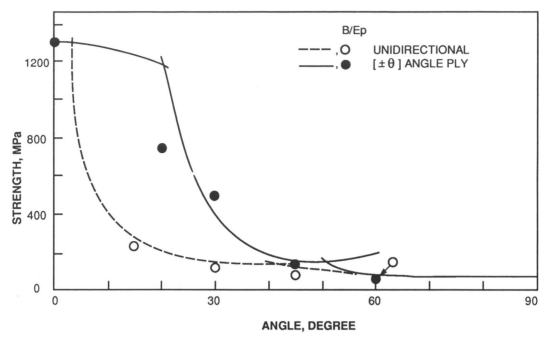

FIGURE 1.4-21. Comparison between angle-ply laminates and off-axis unidirectional laminas.

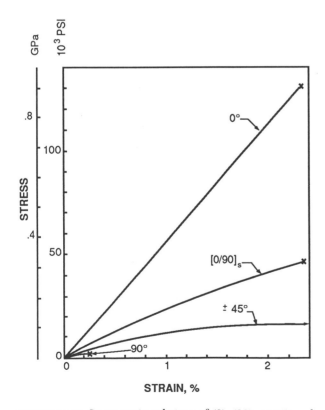

FIGURE 1.4-22. Stress-strain relations of (0), (90), (±45) and [0/90]$_s$ Gl/Ep laminates respectively.

For angles greater than 45°, the difference between the two types of laminates is negligible, and the ply failure is mostly caused by transverse tension. Below 45°, however, the angle-ply laminates are stronger than the unidirectional ones because part of the transverse and shear stresses in the plies are now transferred to the fibers through interlaminar stresses [76].

When an angle-ply laminate is loaded in tension, the first ply failure (FPF) occurs in both $+\theta$ and $-\theta$ plies. That is, cracks are formed parallel to fibers at the weakest locations. These cracks, however, do not lead to an immediate failure of the laminate because further crack propagation through the thickness is prevented by the neighboring plies with the opposite fiber direction. As the load is increased further, more cracks appear in the plies. Eventually, cracks in different plies are connected through delamination, and the final failure ensues.

Whereas both $+\theta$ and $-\theta$ plies fail concurrently in angle-ply laminates and these ply failures lead to the laminate failure soon after, such is not the case in a laminate where loading is in one of the fiber directions. To illustrate this point further, consider a [0/90]$_s$ Gl/Ep laminate subjected to tension in the 0-degree direction. The failure process of this laminate can be understood with the help of the uniaxial stress-strain relations of (0) and (90) laminae in Figure 1.4-22.

The constituent plies in the laminate are not in a state of uniaxial tension although the laminate is. However, the state of stress in each ply is such that the in situ failure strains of the plies are not expected to be much different from the uniaxial failure strains shown in the figure. Therefore, the 90-degree ply will fail first, i.e., the first ply failure occurs in the 90-degree ply, when the composite strain reaches the transverse failure strain [5,7,77–84]. Physically, a normal crack appears parallel to the fibers across the thickness of the 90-degree ply at the weakest location. As the load is increased, there are three possible modes of further damage growth, as schematically shown in Figure 1.4-23.

If both the interface and the neighboring 0-degree plies are strong, the crack cannot grow any further, and another crack will be formed in the 90-degree ply away from the first one [Figure 1.4-23(a)]. If the interface is weak, however, the crack will grow along the interface, producing delamination [Figure 1.4-23(b)]. Finally, a strong interface and sufficiently high stress concentration make possible further propagation of the crack into the 0-degree plies [Figure 1.4-23(c)]. In reality, all three modes are observed with one mode prevailing over the others depending on the stress level. The actual sequence of failure is the first ply failure, the crack multiplication (a), the delamination (b), and finally failure of the 0-degree plies (c).

The crack density (per unit length) in the 90-degree plies of $[0_2/90_2]_s$ Gl/Ep is shown in Figure 1.4-24 as a function of applied stress [82]. The first ply failure stress N_{FPF}/h is calculated from the laminated plate theory and the transverse tensile strength of unidirectional Gl/Ep. The crack density rapidly increases initially and then levels off to a plateau as the applied stress approaches the final failure stress. The predicted crack density at the FPF stress is based on the average transverse stress in 90-degree plies reaching a maximum at a distance from the first crack. Once the first generation of cracks is formed, the next generation occurs between two adjacent cracks. Therefore, the total number of cracks doubles whenever new cracks are formed. However, the actual crack multiplication is much smoother and less than predicted, the difference increasing with the applied stress. The reason is that, at high applied stresses, delamination begins to take place relaxing the ply stress in the 90-degree plies. The figure also shows permanent strains resulting from the cracks.

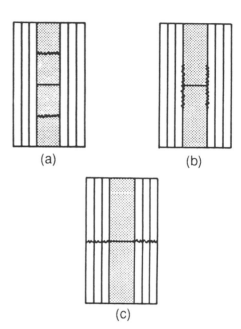

FIGURE 1.4-23. Possible modes of damage growth observed on edge of $[0/90/0]_T$ laminate.

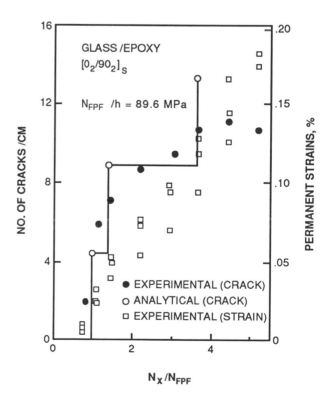

FIGURE 1.4-24. Comparison of measured and analytically predicted crack density in the 90-degree plies of $[0_2/90_2]_s$ Gl/Ep laminate subjected to static tension.

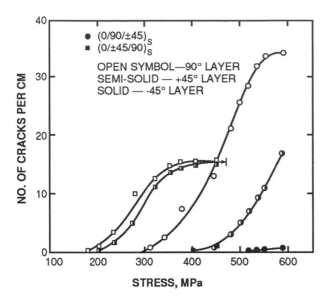

FIGURE 1.4-25. Crack densities in plies of quasi-isotropic Gr/Ep laminates in static tension.

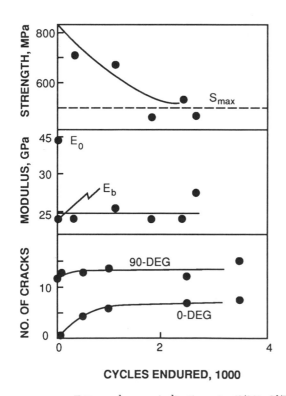

FIGURE 1.4-26. Fatigue damage indications in (0/90) Gl/Ep laminate.

Not only the 90-degree plies but also the 0-degree plies are susceptible to cracking before final failure. According to the laminated plate theory, the transverse stress in the 0-degree plies can be as high as 6 percent of the applied stress. Therefore, in a cross-ply S-Gl/Ep laminate whose strength is as high as 830 MPa, longitudinal cracks can occur in the 0-degree plies because the transverse strength is only 35 MPa [78].

The failure modes observed in (±45) and (0/90) laminates also prevail in (0/±45/90) laminates. The uniaxial stress-strain curve of a (±45) Gl/Ep laminate is shown in Figure 1.4-22. Note that the failure strain of the (±45) laminate is larger than that of the (0) lamina.

When a (0/±45/90) laminate is loaded in uniaxial tension, the 90-degree plies fail first. The cracks in these plies then lead to cracking of the neighboring 45-degree plies as shown in Figure 1.4-20. Upon further increase of load, delamination between plies can precede final failure of the laminate.

Figure 1.4-25 shows the crack densities in plies of two quasi-isotropic Gr/Ep laminates, observed on an edge in static tension [85]. The crack multiplication behavior differs from ply to ply and depends on stacking sequence. The initial crack multiplication in these laminates is much more gradual than in the (0/90) Gl/Ep laminate of Figure 1.4-24.

The failure processes in tension dominated fatigue of composite laminates are similar to those in static tension. An analogy can be drawn between the two if the applied stress in static loading is replaced by the number of cycles endured in fatigue. A major difference is that in fatigue, much more subcritical damage growth precedes final failure, a manifestation of higher fatigue sensitivity of the matrix and interface [85].

A laminate may not delaminate in static tension, but it can suffer extensive delamination in fatigue [5,82,85]. The maximum number of cracks in plies before final failure is larger in fatigue than in static tension [5,85]. Also, off-axis plies, such as ±45-degree plies in a (0/±45) laminate, which do not fail in static tension, exhibit gradual cracking in fatigue. The amount of delamination is not easily correlated with fatigue life: at the same stress level, specimens of shorter life may show less delamination than specimens of longer life [86].

Figure 1.4-26 shows the number of cracks in plies of a (0/90) Gl/Ep laminate versus the number of fatigue cycles. The cracks were counted within a preselected field of view [78]. As the fatigue stress is well above the FPF stress, many cracks appear in the 90-degree plies even after the first cycle. Under constant strain amplitude testing, when a new crack appears between any two adjacent cracks in the same ply, the maximum ply stress is reduced substantially. Therefore, it takes much longer for the next generation of cracks to be formed. The resulting crack multiplication is thus asymptotic with fatigue cycles. Note the crack multiplication in the 0-degree plies. It should be kept in mind that delamination reduces the possibility of further cracking of plies.

The changes in modulus and strength caused by the ply failures are also shown in the figure. These changes will be discussed later in detail.

A similar crack multiplication behavior in fatigue of $[0/90/\pm45]_s$ Gr/Ep is shown in Figure 1.4-27 [85]. Here all data at several fatigue stresses have been combined using the cycle ratio and the damage ratio. The damage ratio is the ratio of the crack density at n cycles to the crack density at final failure. Figure 1.4-28 shows that the crack density at final failure is fairly independent of fatigue stress unless the fatigue stress is above a threshold level [85]. Therefore, the constant crack density below the threshold level is called the equilibrium crack density. Above the threshold stress, the crack multiplication is not fast enough to reach the equilibrium crack density before final failure. However, below the threshold stress, the crack density reaches the equilibrium value well before final failure.

Unlike homogeneous materials, polymer-matrix composites do not, at least yet, seem to render themselves amenable to post-mortem failure analysis through microscopic fractography. Even when an off-axis unidirectional lamina fails in fatigue, it seems to do so rather suddenly without leaving any evidence of crack nucleation and stable growth [71]. When the failure of unidirectional Gr/Ep is caused by transverse tensile stress, the fracture is smooth on the matrix-fiber level and the matrix shows a cleavage type of failure. On the other hand, high longitudinal shear stress gives rise to many matrix cracks only one fiber-diameter deep normal to fibers [69,71]. However, the fatigue fracture surfaces are hardly distinguishable from the static fracture surfaces [71].

The matrix cracks are almost always normal to fibers on any fracture surface. Figure 1.4-29 shows a delamination surface of $[0/\pm45]_s$ Gr/Ep. Note the turning of the crack direction as the fiber orientation changes. In

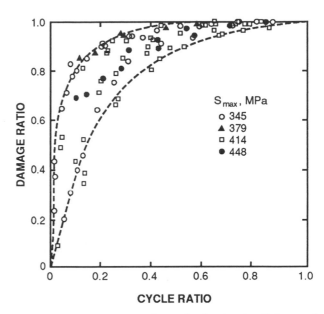

FIGURE 1.4-27. Increase of crack density in fatigue of $[0/90/\pm45]_s$ Gr/Ep.

some cases, the delamination surface consists of two areas of different shades: one shiny and the other dull [86]. The shiny area contains bare fibers while the dull area shows only fiber imprints. Apparently, the bare fibers act as convex mirrors reflecting out more light. The fiber imprints are like concave mirrors, giving the surface a dull appearance.

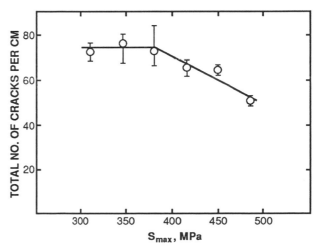

FIGURE 1.4-28. Effect of fatigue stress on crack density at final failure of $[0/90/\pm45]_s$ Gr/Ep.

FIGURE 1.4-29. Delamination surface of $[0/\pm 45]_s$ Gr/Ep.

So far we have discussed general failure processes in composite laminates. As briefly mentioned earlier, the ply failures and delamination lead to the change of properties. In the following, therefore, we shall discuss the effects of these subcritical failures on mechanical properties. Since the failure processes no doubt depend on many parameters, such as stacking sequence, loading condition, and environment, these parameters will be discussed also.

Stress-Life (S-N) Relations

S-N relations of various B/Ep laminates are shown in Figure 1.4-30 [34,63]. The ratio of maximum fatigue stress to average ultimate tensile strength has been used on the ordinate. All the curves except for the two quasi-isotropic laminates are essentially within the same band. Thus, the fatigue sensitivity of B/Ep laminates is fairly independent of laminate type and similar to that of unidirectional composites, as long as these laminates contain a sufficient percentage of 0-degree plies.

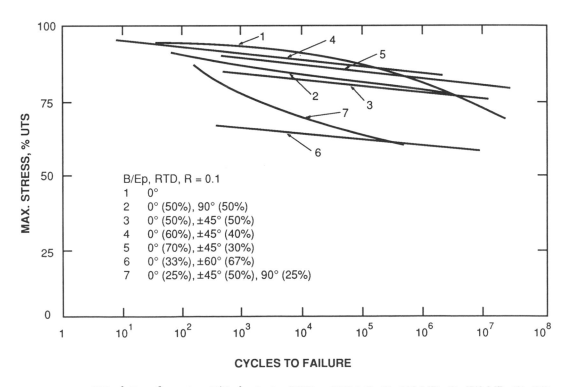

FIGURE 1.4-30. *S-N* relations for various B/Ep laminates [UTS = 1331 MPa (1), 623 MPa (2), 752 MPa (3), 779 MPa (4), 972 MPa (5), 427 MPa (6), 338 MPa (7)].

The large reduction in fatigue strength of quasi-isotropic laminates 6 and 7 is probably the result of decreasing load sharing in fatigue by the off-axis plies. Since these laminates contain relatively small percentages of 0-degree plies, degradation of the off-axis plies will be more noticeably reflected on the fatigue strength of the laminates.

To check the applicability of the rule of mixtures to static and fatigue strengths, the laminate-to-lamina strength ratio is plotted against the fraction of 0-degree plies for the same laminates (Figure 1.4-31). The rule of mixtures is seen to provide a reasonable estimate of laminate strengths, both static and fatigue, for the laminates other than the quasi-isotropic laminates. For the latter laminates, the actual fatigue strengths are lower than predicted.

The off-axis plies in Gr/Ep laminates seem more effective in load sharing than those in the B/Ep laminates, as illustrated in Figure 1.4-32 [87]. In particular, the quasi-isotropic Gr/Ep laminates are considerably stronger than predicted by the rule of mixtures based on the strength of the 0-degree ply alone. The data below the rule of mixtures line indicate that off-axis plies are not always beneficial, probably because cracks in these plies cause stress concentrations on the load-carrying 0-degree plies. In both B/Ep and Gr/Ep, (0/90) laminates fall below the rule of mixtures predictions.

The effect of off-axis plies on the fatigue ratio of Gr/Ep laminates is shown in Figure 1.4-33. Again, the quasi-isotropic laminate has the lowest fatigue ratio. However, it is about the same as the fatigue ratio of unidirectional composites.

Normalized *S-N* relations are shown for quasi-isotropic laminates of B/Ep, Gr/Ep and Gl/Ep, respectively, in Figure 1.4-34 [50,63,88]. The Gr/Ep laminate is seen to be most fatigue resistant, followed by the B/Ep laminate. Most susceptible to fatigue is the Gl/Ep laminate. In the figure both B/Ep and Gl/Ep laminates have a stacking sequence of $[0/\pm45/90]_s$ whereas the Gr/Ep laminate was of $[0/45/90/-45_2/90/0]_s$. It is not clear whether the difference in stacking sequence may be responsible for the different fatigue sensitivity between the B/Ep and Gr/Ep laminates.

The scatter in fatigue life can be analyzed by using the Weibull distribution, Equation (1.4-1). Figure 1.4-35 shows the shape parameter increasing with decreasing fatigue stress for Gr/Ep laminates [86,89,90]. Thus, the underlying failure process seems to be more

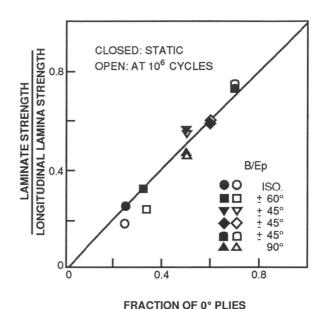

FRACTION OF 0° PLIES

FIGURE 1.4-31. Ratio of laminate strength to lamina strength versus fraction of 0-degree plies for the laminates of Figure 1.4-30. Both static and fatigue strengths are shown. Numbers represent fiber directions of off-axis plies.

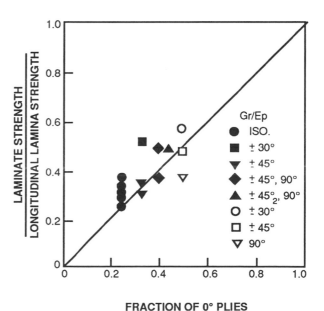

FRACTION OF 0° PLIES

FIGURE 1.4-32. Ratio of laminate strength to lamina strength versus fraction of 0-degree plies for Gr/Ep laminates in static tension.

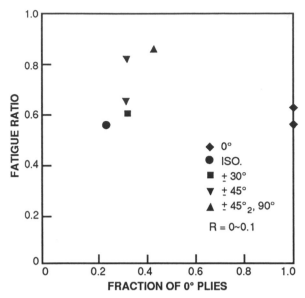

FIGURE 1.4-33. Fatigue ratios (at 10^6 cycles) of fiber-controlled Gr/Ep laminates.

of a wear-out type at lower stresses. However, it is not known what the shape parameters would be at still lower stresses. Perhaps a true fatigue limit may exist near 50% UTS.

Effect of Compression

Composite laminates are weaker in compression-dominated fatigue than in tension-dominated fatigue because the subcritical failures discussed earlier are

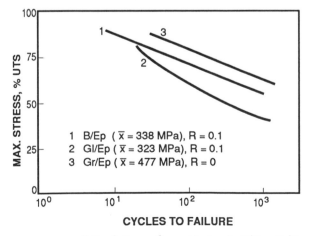

FIGURE 1.4-34. S-N relations of quasi-isotropic B/Ep, Gr/Ep and Gl/Ep laminates, respectively ($\bar{x}$ = UTS).

more deleterious to compression than to tension. Even in static loading, unidirectional Gl/Ep and Kv/Ep composites are much weaker in compression (see section 1.3).

A constant-life diagram of $[0_2/\pm45/0_2/\pm45/90]_s$ Gr/Ep is shown in Figure 1.4-36 [91]. Since the static strength is the same in tension as in compression, the static strength curve is symmetric with respect to the alternating stress axis. However, with increasing life the constant-life curve becomes more and more skewed to the right. That is, the peak of constant-life curve moves into the tension-dominated quadrant. A similar behavior is observed in other Gr/Ep laminates [8,92].

When a laminate statically is stronger in compression than in tension, as in B/Ep composites, the peaks of the constant-life curves in the low-cycle region are in the compression-dominated quadrant [10,34,63]. In the high-cycle region, however, the peaks again occur in the tension-dominated quadrant.

Change in Modulus

The stress-strain relation of $[0/\pm45/90]_s$ Gl/Ep shown in Figure 1.4-22 is linear up to the FPF point. Above the FPF point, the slope decreases with increasing stress up to final failure. If the laminate is unloaded from above the FPF point, the unloading curve shows a smaller slope than the initial tangent slope because of failure of off-axis plies. Since the number of cracks in off-axis plies increases with fatigue, the modulus will decrease.

The cracks in plies produce some permanent strain, as seen in Figure 1.4-24. If the permanent strain is neglected, the loading-unloading curves in fatigue can be idealized as shown in Figure 1.4-37. In the figure E_0 is the initial tangent modulus and E_{sn} is the secant modulus to the maximum fatigue stress after n cycles.

Figure 1.4-38 shows the change of secant modulus in fatigue of a $[0/\pm45/90]_s$ Gl/Ep laminate at various stress levels. Also shown are the unidirectional lamina data, which show no change in modulus until part of the specimen is cut off by longitudinal cracks. Thus, the change in secant modulus of the laminate is caused mostly by failure of off-axis plies and delamination, as suggested earlier.

The symbols on the ordinate represent the secant moduli after the first cycle. As might be expected from Figure 1.4-37, this initial reduction increases with fatigue stress level.

In stiffness-critical design, due consideration should be given to any change of stiffness in fatigue. On the other hand, the change of stiffness can be taken as a measure of fatigue damage. It is interesting to observe in Figure 1.4-38 that the secant modulus at the end of fatigue life is close to the secant modulus at static failure.

The secant modulus in fatigue of [0/90/±45]$_s$ Gr/Ep laminate is shown in Figure 1.4-39. This laminate is the same as the one used in Figures 1.4-27 and 1.4-28. The modulus reduction is greater at the lower stress level. This observation, coupled with the larger final crack density at the lower stress, again indicates that the modulus reduction is mainly the result of ply failures and delamination.

Change in Strength

Whereas unidirectional composites do not show much change in strength until immediately before final failure, laminates undergo a gradual strength reduction in fatigue. In (0/90) Gl/Ep the strength reduction is rapid initially and then slows down as fatigue proceeds further [78,80,93]. Final fracture occurs rather suddenly.

Figure 1.4-40 shows residual tensile and compressive strengths after fatigue of the quasi-isotropic Gr/Ep laminate in Figure 1.4-34 [88]. Initially, the tensile and compressive strengths were equal to each other. However, in fatigue the compressive strength decreases faster than the tensile strength, the difference increasing with fatigue cycles. The reason is that ply failures and delamination are more detrimental in compression than in tension, making compression buckling of 0-degree plies easier. Note also that the cycle ratio has less effect than the number of cycles itself.

Temperature Increase

Cyclic loading of a laminate with ply failures and delamination combined with a viscoelastic matrix generates heat within the laminate [94]. A typical temperature increase on the specimen surface consists of three stages (Figure 1.4-41). The first stage of rapid temperature increase is merely a result of transient heat transfer between specimen and environment. The second stage of constant rate of temperature increase is directly related to the rate of heat dissipation within the specimen. If the temperature increase during this stage has a

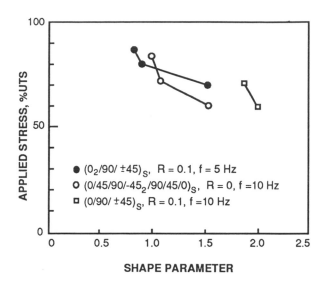

FIGURE 1.4-35. Change of fatigue shape parameter with fatigue stress.

positive slope, the rate of heat dissipation increases with fatigue. The accelerated temperature increase in the third stage is a manifestation of extensive damage occurring before final failure [95–97].

In many cases of moderate fatigue loading, the temperature in the second stage remains fairly constant, indicating a constant rate of heat dissipation. The corresponding equilibrium temperature increase ΔT_e can be predicted from a heat transfer analysis under the assumption of a constant rate of heat generation proportional to the area of hysteresis loop in the stress-strain diagram in fatigue [50,95]. Specifically, the equilibrium temperature increase ΔT_e is related to the loading parameters by:

$$\Delta T_e \propto (1 - R)^2 S_{max}(S_{max} - S_0)f \qquad (1.4\text{-}10)$$

where S_0 is the threshold stress below which no hysteresis is observed and f is the test frequency [50]. Figure 1.4-42 shows experimental equilibrium temperature increases in [0/±45/90]$_s$.

A positive slope of the second-stage temperature increase is an indication of damage rate increasing with fatigue cycles. In polymers this type of increase is associated with thermal fatigue [98]. If such an increase is prevented by artificial cooling, a substantial increase in life can result [97].

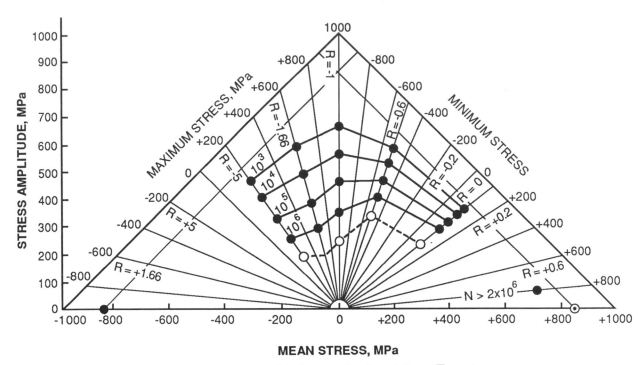

FIGURE 1.4-36. Constant life diagram of $[0_2/\pm 45/0_2/\pm 45/\overline{90}]_s$, Gr/Ep.

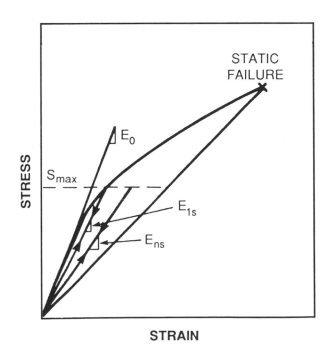

FIGURE 1.4-37. Idealized loading-unloading curves for a laminate.

The temperature distribution on the specimen surface may not be uniform even in a smooth specimen if the internal damage is not uniformly distributed. Such nonuniform temperature distribution can be used to locate internal defects [67]. Naturally, the temperature increase at stress raisers, such as crack tips and hole boundaries, is much higher because of the higher strain energy density [94,99,100].

Frequency Effect

An increased test frequency usually leads to a higher number of cycles to failure if the slope of the second-stage temperature increase remains zero [10,95,96]. Otherwise, a shorter life results because of an accelerated damage process. In the case of the $[0/\pm 45]_s$ Gr/Ep [10] results shown in Figure 1.4-43, the test frequency has minimal effect on fatigue life. At every frequency tested, an equilibrium temperature increase was observed throughout almost the entire fatigue life.

Since matrix materials are sensitive to loading rate, matrix-controlled properties are expected to be susceptible to the test frequency [101].

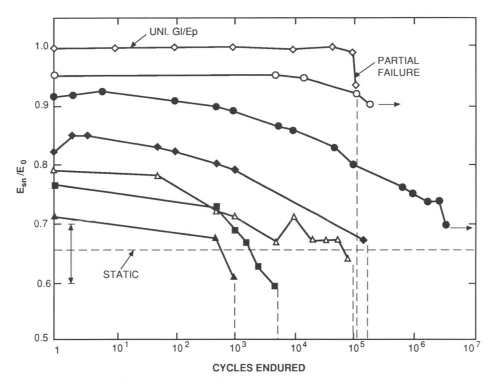

FIGURE 1.4-38. Change of secant modulus in fatigue of $[0/\pm45/90]_s$ Gl/Ep laminate.

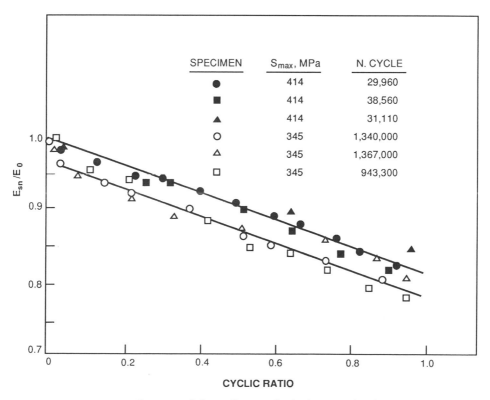

FIGURE 1.4-39. Secant modulus in fatigue of $[0/90/\pm45]_s$ Gr/Ep laminate.

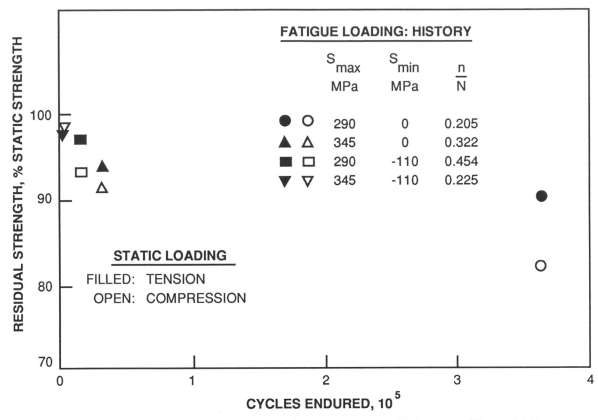

FIGURE 1.4-40. Strength reduction in fatigue of the quasi-isotropic Gr/Ep laminate of Figure 1.4-34.

Whether fatigue failure is a result of sustained loading or of repeated loading-unloading was investigated through interrupted stress rupture tests of a Gr/Ep laminate [102]. It was found that the number of cycles to failure increased with frequency in a manner that the total time under load remains constant.

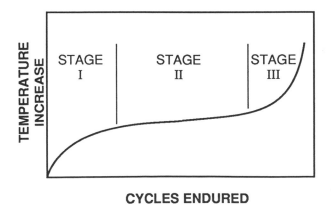

FIGURE 1.4-41. Typical temperature increase.

Fatigue Notch Sensitivity

One of the biggest advantages of composites over metals is their very low fatigue notch sensitivity. In static tension, a $[0/\pm45/90]_s$ Gr/Ep laminate is more sensitive to a notch than a 7075 Al plate of comparable thickness. However, in fatigue, notched laminates in essence have built-in crack arrestors. Consequently, fatigue notch sensitivity is lower than static notch sensitivity.

Figure 1.4-44 shows the inverse of fatigue notch factor versus fatigue cycles for B/Ep laminates with notches [34]. Note that the fatigue notch sensitivity decreases with fatigue cycles.

The excellent resistance to fatigue in the presence of notches can be explained in terms of the damage modes at notch tips. Figure 1.4-45 schematically compares static and fatigue damages observed at crack tips in a $(0/\pm45)$ Gr/Ep laminate. Actual damages were radiographed after applying tetrabromoethane to enhance the images [103,104]. In the static case, as much cracking occurs in the ±45-degree plies as in the 0-degree plies. However, in fatigue cracks grow much longer in the 0-degree plies, thus relieving much of the stress concentration on the 0-degree plies caused by the notch. Therefore, the 0-degree plies will behave as if there were no crack. The result can be a residual strength higher than the static notched strength [104–106]. A similar behavior is also observed for holes [92, 107–110].

Notches may be no more deleterious in compression than in tension. Figure 1.4-46 shows a constant life diagram for the same lay-up as in Figure 1.4-36, with a hole [111]. The constant life curve at 10^6 cycles is almost symmetric with respect to the ordinate, indicating as good a resistance to compression-dominated fatigue as to tension-dominated fatigue. Other available data also indicate good resistance to compression-dominated fatigue in the presence of a hole [112,113]. Further results are provided in references [114,115].

Unlike homogeneous materials, composite laminates do not allow self-similar crack growth except in a few special cases [116]. A method of predicting fatigue life in the presence of a notch has been proposed in reference [117].

Effect of Stacking Sequence

So far the discussion has essentially been within the framework of the classical laminated plate theory [118], and no distinction has been made between laminates with different stacking sequences. However, failure behavior, especially in fatigue, of laminates depends very much on stacking sequence.

The effect of stacking sequence on damage development in quasi-isotropic Gr/Ep laminates was studied in references [82,85]. Figure 1.4-25 shows that the $[0/90/\pm45]_s$ laminate has more cracks in its constituent plies than the $[0/\pm45/90]_s$ laminate does at the same stress level because the 90-degree plies in the second laminate are lumped together. Since the crack spacing increases with the thickness of the plies in which cracks are located, the first laminate will have more cracks than the second.

Figure 1.4-25 also shows the FPF stress of the $[0/90/\pm45]_s$ laminate being higher than that of the $[0/\pm45/90]_s$ laminate. In fact, the FPF stress was found to increase with decreasing thickness of lumped 90-degree plies in laminates [85,119]. This behavior was explained by assuming that the failure of the 90-degree plies was initiated at an inherent crack.

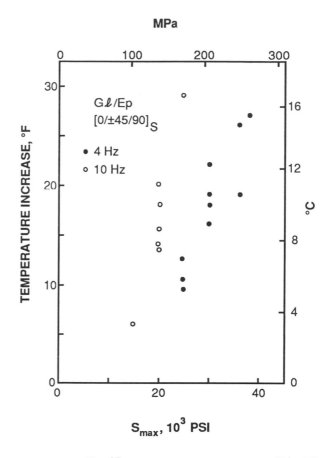

FIGURE 1.4-42. Equilibrium temperature increase in $[0/\pm45/90]_s$ Gl/Ep laminate.

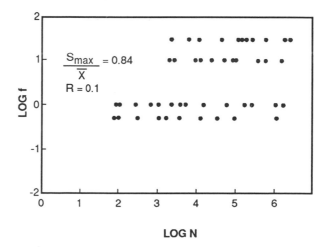

FIGURE 1.4-43. Effect of test frequency on fatigue life of $[0/\pm45]_s$ Gr/Ep.

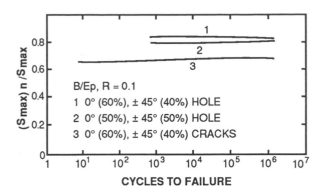

FIGURE 1.4-44. Inverse of fatigue notch factor for B/Ep laminates.

In fatigue, more cracks develop before final failure than is the case in static tension, as shown in Figure 1.4-47 [85]. Still, the $[0/90/\pm45]_s$ laminate contains more cracks than the $[0/\pm45/90]_s$ laminate.

Another example of a stacking sequence effect is delamination along free edges. As the result of many investigations [76,120,121], it is well-known that free edges are subjected to very high interlaminar stresses. The free-edge interlaminar stress in the $[0/\pm45/90]_s$ laminate is tensile while that in the $[0/90/\pm45]_s$ laminate is compressive. Therefore, the $[0/\pm45/90]_s$ laminate is much more susceptible to free-edge delamination in tension-dominated fatigue [85]. More recently O'Brien et al. [122] developed a free-edge delamination

specimen that has been employed in cyclic delamination growth investigations [123,124].

Environmental Effect

The anisotropy in hygrothermal expansion behavior of unidirectional composites manifests itself in residual stresses within the laminate [125,126]. In a room environment after fabrication, laminates are subjected to curing stresses caused by thermal expansion mismatch between plies. Therefore, an environmental change not only affects intrinsic material properties, but also the magnitude of residual stress, which in turn affects the mechanical behavior of laminates.

Effects of temperature and moisture on static properties are well-summarized in reference [127]. The results indicate that, as expected, fiber-controlled properties are much less affected than matrix-controlled properties. Absorbed moisture lowers the glass transition temperature of polymer matrix [128]. In Gl/Ep laminates, absorbed moisture degrades the fiber-matrix interface and contributes to stress corrosion of glass fibers [129].

Figure 1.4-48 shows effects of moisture and temperature on tensile fatigue strength of a fiber-controlled $(0_3/\pm45)$ B/Ep laminate [35]. In this figure, RTD denotes "room temperature, dry" and RTW denotes "room temperature, wet." Both moisture and temperature increase fatigue sensitivity. The corresponding static strength reductions are shown on the ordinate. The static strength reduction at 177°C has been estimated from the data for a (0/90) laminate. In some cases, however, moisture may have a beneficial effect on fatigue strength [5,130].

The hygrothermal effects on fatigue of Gr/Ep laminates are similar to those for B/Ep laminates [67]. However, water can reduce the fatigue limit of fiberglass cloth reinforced polyester by as much as 30 percent [131].

Other Loading Parameters

The available data on the effects on fatigue life of loading parameters, such as overloading, spectrum loading and biaxial loading, are still limited. Interested readers are referred to references [42,86,90,132] for proof testing, to references [22,91,133] for spectrum loading, and to references [134,135] for biaxial loading. Various statistical treatments of fatigue failure are found in references [10,136–138].

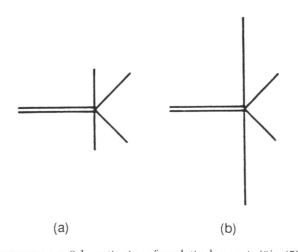

(a)   (b)

FIGURE 1.4-45. Schematic view of crack-tip damage in $[0/\pm45]$ Gr/Ep: (a) static tension; (b) fatigue.

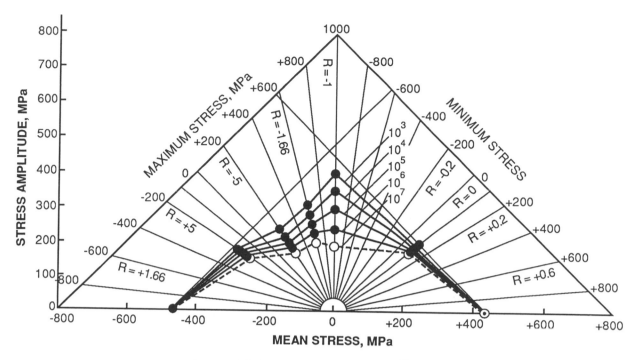

FIGURE 1.4-46. Constant life diagram for $[0_2/\pm 45/0_2/\pm 45/\overline{90}]$, Gr/Ep with a hole.

1.4.4 Discontinuous Fiber Composites

The load in discontinuous fiber composites (DFC) is transferred from fiber to fiber through the matrix and interface. The stress concentrations at fiber ends may accelerate crack initiation. However, further crack growth is retarded by fibers. If fibers are relatively long as in sheet molding compound (SMC) composites, they can still bridge matrix cracks, thus allowing multiple cracking before final failure. As fibers become shorter, cracks can grow around fibers or fibers can more easily be pulled out. Consequently, not much multiple cracking is observed.

Thus, the fatigue sensitivity of DFC is the net result of crack initiation, crack growth, and multiple cracking. The multiple cracking, which distinguishes composites from homogeneous materials, increases with fiber length. However, longer fiber length accelerates crack growth along the fiber. Since crack growth is more frequently retarded by the random arrangement of fibers, the cracks in discontinuous fiber composites are much shorter and more shallow than those in continuous fiber composites.

In this subsection we shall first describe the stress

concentration at a fiber end. We shall then discuss failure processes and stress-life relations. The effects of fatigue damage on mechanical properties will also be included. The composites to be discussed are chopped strand mat, SMC, and injection molded fiber reinforced thermoplastics.

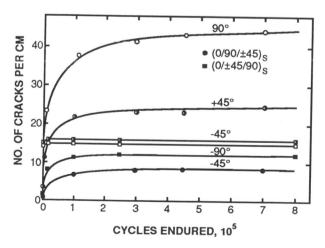

FIGURE 1.4-47. Crack densities in fatigue of the laminates of Figure 1.4-25.

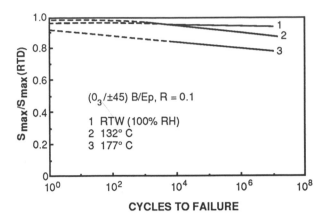

FIGURE 1.4-48. Effects of moisture and temperature on fatigue strength of $[0_3/\pm 45]$ B/Ep.

State of Stress Near a Fiber End

Due to the presence of many fiber ends, the state of stress in discontinuous fiber composites (DFC) is much more complex than in continuous fiber composites (CFC). As a result much effort, both analytical and experimental, has been directed toward the determination of stresses around a fiber of finite length embedded in a composite.

An idealized situation of a fiber in a composite is shown in Figure 1.4-49. Here a fiber of length ℓ is surrounded by the matrix, and the composite outside this matrix is regarded as a homogeneous continuum. One of the simplest analysis models is the so-called shear lag model. There are several variations of this model

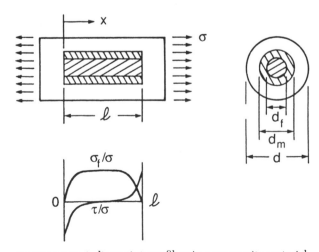

FIGURE 1.4-49. A discontinuous fiber in a composite material.

[139–141]. Here we shall briefly describe the results given by Rosen [141] in order to qualitatively demonstrate discontinuous fiber effects on the stresses in a DFC.

The equilibrium condition for the fiber in the axial direction is expressed by:

$$\frac{d_f}{4}\frac{d\sigma_f}{dx} + \tau = 0 \tag{1.4-11}$$

where σ_f is the average axial stress in the fiber, τ is the shear stress in the matrix, and d_f is the fiber diameter. For the entire composite to be in equilibrium, the average stress on any cross section should be equal to the applied stress σ:

$$d_f^2\sigma_f + (d^2 - d_m^2)\sigma_c = d^2\sigma \tag{1.4-12}$$

where σ_c is the axial stress in the outside composite, d_m is the characteristic dimension of the matrix region, and d is the characteristic dimension of the composite. Note that in this treatment no axial stress is present in the matrix. Expressing the matrix shear strain in terms of the fiber displacement and the outside composite displacement, and using the constitutive relations, we can derive the third equation necessary to complete the formulation:

$$\frac{d_m - d_f}{2}\frac{1}{G_m}\frac{d\tau}{dx} = \frac{1}{E}\sigma_c - \frac{1}{E_f}\sigma_f \tag{1.4-13}$$

where E is the composite modulus, E_f the fiber modulus, and G_m the matrix shear modulus. Assuming no bond at the fiber end and noting that $d \gg d_m$ in actual composites, we finally obtain the solution:

$$\frac{\sigma_f}{\sigma} = \frac{E_f}{E}\left(1 - \frac{\cosh \beta(\bar{\ell} - 2\bar{x})}{\cosh \beta\bar{\ell}}\right) \tag{1.4-14}$$

$$\frac{\tau}{\sigma} = -\frac{E_f}{E}\frac{\beta}{2}\frac{\sinh \beta(\bar{\ell} - 2\bar{x})}{\cosh \beta\bar{\ell}} \tag{1.4-15}$$

Here β is defined in terms of the fiber volume fraction $\dot{V}_f$ and the constituent moduli:

$$\beta = \left(\frac{2\sqrt{V_f}}{1 - \sqrt{V_f}}\frac{G_m}{E_f}\right)^{1/2} \tag{1.4-16}$$

and $\bar{\ell}$ and $\bar{x}$ are the normalized fiber length and coordinate, respectively:

$$\bar{\ell} = \ell/d_f, \quad \bar{x} = x/d_f \qquad (1.4\text{-}17)$$

The stresses σ_f and τ are schematically shown in Figure 1.4-49. The high concentration of shear stress at the fiber ends portends the possibility of matrix failure or debonding at the interface even when the fiber is parallel to the loading direction. The increased load sharing by matrix and interface manifests itself in the lower strength of discontinuous fiber composites.

Failure Processes

The damage initiation in discontinuous fiber composites is similar to that in continuous fiber composites; it takes the form of matrix/interface cracking along the fibers normal to the tensile load. Additionally, however, cracks in DFC can be initiated from fiber ends because of high stress concentrations there. Yet the DFC cracks, compared with those in continuous fiber composites, are kept smaller in length and depth by the random arrangement of fibers.

Fatigue of discontinuous fiber composites has been studied by Owen et al. [142–144] and Mandell et al. [145–147] among others; much of this subsection on discontinuous fiber composites is based on their findings.

In composites reinforced with chopped strand mats of glass, fatigue damage initiation is in the form of fiber-matrix debonding followed by resin cracking [142]. The debonding occurs at the ends of the fibers parallel to the loading direction and also along the fibers normal to the loading, the latter mode being the typical damage initiation mode in continuous fiber composites, as discussed in the preceding subsections.

The addition of flexible resin may not increase the strain much at the onset of debonding. In fact, Smith and Owen [148] reported that various glass reinforced laminates exhibited debonding at about 0.3% strain almost regardless of the resin type when the reinforcement was in the form of chopped strand mat, woven cloth, or nonwoven cross-ply. However, Christensen and Rinde [149] showed that the transverse tensile failure strain could be increased from 0.24% to as much as 1% by using flexible epoxies.

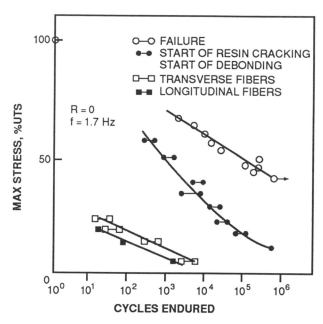

FIGURE 1.4-50. Effect of stress on fatigue damage initiation for chopped strand mat-polyester resin laminates with 30 wt% of flexibilizer added [148].

In fatigue, debonding occurs at much lower strains than in static tension. Figure 1.4-50 shows that the debonding stress in fatigue at 10^3 cycles is only $\sim 40\%$ of the debonding stress in static tension although no final failure is expected well beyond 10^6 cycles. Thus, a design based on debonding will not make DFC competitive with other materials.

The number of debonded fibers and the total lengths of matrix cracks observed at the specimen edges increase almost the same way as does the ply crack density discussed for CFC in subsection 1.4.3. The main difference is that the debonding damage in DFC increases rapidly near final failure whereas both the ply crack density and the matrix crack length increase asymptotically until final failure [142].

The cracking in sheet molding compound (SMC) composites occurs mostly in resin rich regions parallel to the local fibers normal to the loading [145]. In SMC-R25 the first damage in either static tension or fatigue is matrix cracking rather than debonding. However, the fiber-matrix debonding prevails as the first damage mode in SMC-R55. Debonding damage is later followed by matrix cracking in fatigue. Nevertheless, the strain at the first damage in static tension is $\sim 0.3\%$ for both composites.

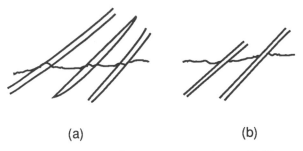

FIGURE 1.4-51. Fibers bridging a matrix crack in an SMC composite: (a) broken fibers; (b) unbroken fibers [145].

Whereas the ply cracks in continuous fiber composites are across the entire specimen width and ply thickness, the cracks in SMC are quite limited in length and depth because of the random fiber arrangement [150]. In most cases matrix cracks terminate when fibers cross their path. Some cracks can propagate around fibers if these fibers are not in tight strands. Although most fibers bridging a crack are not broken, some fibers do fracture, especially in R50 composites after many fatigue cycles. The two types of interaction between fibers and a matrix crack are schematically shown in Figure 1.4-51.

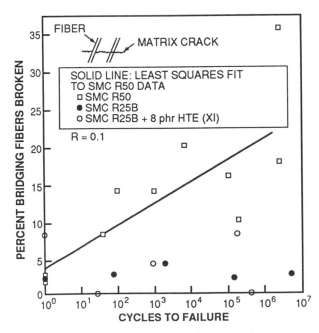

FIGURE 1.4-52. Frequency of breaks of fibers bridging matrix cracks in failed fatigue specimens [145].

The relative number of breaks of fibers bridging matrix cracks in static tension is about 2%, fairly independent of fiber volume content. However, a difference appears in fatigue: the number increases with fatigue lifetime for R50, but it does not seem to change much for R25 (Figure 1.4-52 [142]). The reason for such a difference is not understood at present.

The crack density measured along a line down the center of the specimen surface increases asymptotically in fatigue. There are slightly more cracks observed on the surfaces than in the interior of specimens. The crack density at final failure is fairly independent of the fatigue stress level within the experimental scatter. In fact, the final crack density in static tension is almost the same as that in fatigue (Figure 1.4-53) [145]. The rubberized R25 shows the highest crack density at failure followed by R25 and R50 in that order. At the same fatigue stress level the rubberized R25 is less susceptible to cracking than the unmodified R25. However, the increased toughness of the rubberized matrix allows more cracking without failure. Thus, the higher crack density at failure cannot be taken as a sign of inferior fatigue resistance. The crack densities in Figure 1.4-53 are comparable to the density of 90° ply cracks in the $[0_2/90_2]$ Gl/Ep laminate of Figure 1.4-24.

When the reinforcement fibers are very short, i.e, shorter than 1 mm, as in injection molded thermoplastics, the composites cannot sustain much matrix cracking before final failure [147]. In this case, only a few matrix cracks appear to occur around fiber ends. The crack growing from a notch does so avoiding fibers. Thus, the crack growth mainly consists of matrix cracking and interfacial failure. Also, the probability of fiber fracture by matrix cracks depends on the type of fiber.

Glass fibers are weaker and more susceptible to fracture than carbon fibers. Consequently, carbon fiber reinforcement generally yields higher static and fatigue strengths than glass fiber reinforcement. However, a brittle matrix, e.g., polyphenylene sulfide, combined with glass fibers may result in higher static strength than when combined with carbon fibers [147]. Yet, the brittle matrix/carbon fiber combination leads to improved fatigue resistance.

When the matrix is ductile, the fibers at the fracture surface are loosened by plastic distortion of the matrix surrounding them. Thus, even those fibers that are at an angle to the loading can rotate themselves and be pulled out. However, a brittle matrix does not allow such rotation of fibers although the fibers may debond from the matrix. Thus, complete fracture of the composite requires additional local fracture of the matrix and/or fibers. Therefore, when a brittle matrix is combined with fatigue-resistant carbon fibers, the resulting composite is expected to perform better in fatigue than when it is combined with glass fibers.

To test the foregoing hypothesis of mechanical interlocking of matrix cracks by fibers in polyphenylene sulfide/carbon composites, Mandell et al. [147] compared a fatigue-grown crack with an artificially introduced one as to their tendency to reduce strength. It was found that the composite was much more sensitive to the artificial crack than to the fatigue crack. This finding is in agreement with the foregoing hypothesis because no fibers bridge the artificial crack. Also, the effectiveness of mechanical interlocking was shown to disappear when the matrix was ductile: nylon 66/carbon composite was as sensitive to a fatigue crack as to an artificial crack.

S-N *Relations*

Mandell et al. [146] summarized *S-N* data of various types of glass reinforced composites in the following form:

$$S = UTS - B \log N \qquad (1.4\text{-}18)$$

The composites surveyed are listed in Table 1.4-4. They include unidirectional laminates, chopped strand mat polyester, and several injection molded thermoplastics.

Mandell et al. [146] then plotted UTS versus the slope of *S-N* curve, *B*, as shown in Figure 1.4-54. It is interesting to note that the slope *B* is about 0.1 UTS regardless of the type of composite. They further showed that even the unimpregnated strands had the same *B/UTS* ratio as did the composites. Therefore, they concluded that the fatigue failure of glass reinforced plastics was mainly dominated by glass fibers up to 10^6 cycles at a stress ratio of 0.1.

The *S-N* curves of carbon fiber reinforced thermoplastics were found to be describable by a linear equation of the form of Equation (1.4-18) with their slope depending on the matrix material. Both UTS and *B* of glass and carbon fiber reinforced thermoplastics are listed in Table 1.4-5 for easy comparison. The fibers in these composites were shorter than 0.6 mm [147].

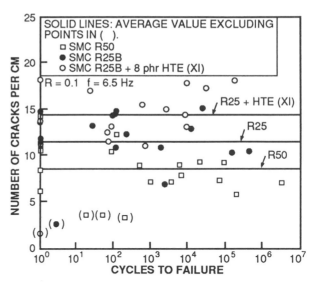

FIGURE 1.4-53. Crack density at failure [145].

Of the five matrix materials in Table 1.4-5, polyphenylene sulfide (PPS) and polyamide-imide (PAI) are brittle. They fractured both in tension and in fatigue. Polysulfone and polycarbonate failed by necking when the fatigue stress was so high that the lifetimes were shorter than $\sim 10^3$ cycles. Nylon 66 is the toughest of all; it never fractured even up to 10^6 cycles which was the maximum cycles applied.

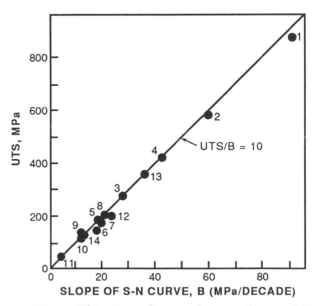

FIGURE 1.4-54. Ultimate tensile strength versus slope on *S-N* curve for various glass reinforced composites ($R = 0$ to 0.1). See Table 1.4-4 for identification of numbers [146].

Table 1.4-4. Composites in Figure 1.4-54 [146].

Composite	No. in Figure 1.4-54	Reference
Unidirectional		
(50 vol% glass)	1	[30]
(33 vol% glass)	2	
(16 vol% glass)	3	
(0/90) Gl/Ep	4	[151]
Nylon 66/Gl	5	[147]
(40 wt% glass)		
Polycarbonate/Gl	6	
(40 wt% glass)		
Polyphenylene Sulfide/Gl	7	
(40 wt% glass)		
Polyamide-imide/Gl	8	
(30 wt% glass)		
Chopped Strand Mat/Polyester	9	[152]
SMC (6.4 mm gage length)	10	[147]
SMC (25 mm gage length)	11	
SMC-R50	12	
[0/±45/90]s Gl/Ep	13	[132]
Chopped Strand Mat/Polyester	14	[153]

In general, the carbon reinforced thermoplastics show lower *B/UTS* ratios than the glass counterparts, indicative of better fatigue resistance. When reinforced with carbon fibers, the brittle matrices PPS and PAI seem to improve their fatigue performance over the glass reinforcement because the fiber rotation through plastic deformation of the matrix is difficult, as discussed in the preceding subsection. The ductile matri-

Table 1.4-5. Strengths and S-N slopes of fiber reinforced thermoplastics [130].

Composite[a]	UTS (MPa)	B (MPa)	B/UTS
Nylon 66/Glass (N66/Gl)	181	19.6	0.11
Polycarbonate/Glass (PC/G)	161	18.2	0.11
Polyphenylene Sulfide/Glass (PPS/G)	180	19.6	0.11
Polyamide-imide/Glass (PAI/G)[b]	202	21.3	0.11
Nylon 66/Carbon (N66/C)	256	25.4	0.10
Polycarbonate/Carbon (PC/C)	203	23.4	0.12
Polysulfone/Carbon (PSUL/C)	197	21.8	0.11
Polyphenylene Sulfide/Carbon (PPS/C)	156	12.2	0.08
Polyamide-imide/Carbon (PAI/C)[b]	231	22.0	0.10

[a]40% by weight of fibers unless otherwise mentioned.
[b]30% by weight of fibers.

ces show only a slight improvement with the carbon reinforcement. The ductility facilitates fiber pull-out even though carbon fibers do not fail easily in fatigue.

The stress-strain relation during stress-controlled fatigue is schematically shown in Figure 1.4-55 [147]. Both the maximum and minimum strains in each cycle increase with fatigue cycles. The cyclic softening indicated by the change of the maximum strain consists of three regions as schematically shown in Figure 1.4-56 [154].

The initial region indicates a transition state and is characterized by a rapid increase of inelastic strain. This region is followed by a steady state region where a gradual softening occurs with a constant rate. The strain rate depends on the material and stress level. The steady state region comprises most of the fatigue process. The final region is characterized by the inelastic strain increasing rapidly until final failure. Below the fatigue limit, however, the strain rate in the final region is almost negligible so that no further change in material is apparent. The curve in Figure 1.4-56, in fact, resembles the creep strain under a constant load. The cumulative strain to failure is the sum of all these strains:

$$\epsilon_{bf} = \epsilon_0 + \epsilon_I + \epsilon_{II} + \epsilon_{III} \qquad (1.4\text{-}19)$$

where

ϵ_0 = initially applied strain
ϵ_I = inelastic strain in initial softening region I
ϵ_{II} = inelastic strain in steady state region II
ϵ_{III} = inelastic strain in final region III

The cumulative strains to failure are shown against the cycles to failure in Figure 1.4-57 for the same fiber reinforced thermoplastics of Table 1.4-5 [147]. The different cycles to failure are the result of different fatigue stresses rather than experimental scatter. For nylon based composites the strain at failure increases initially to ~ 100 cycles and then decreases. Over this transition region the *S-N* curves also turn slightly upward exhibiting a nonlinear behavior.

The increasing cumulative cyclic strain in the low cycle region is an indication that the underlying failure mechanism is one of creep rupture. Note that fatigue stresses are high in the low cycle region. Since nylon 66 is ductile, its failure in the low cycle region is likely to be creep-controlled. Results for nylon 6 are provided in reference [155].

The decreasing cumulative cyclic strain, on the other hand, is characteristic of the failure process controlled by crack growth. Thus, in nylon 66 composites a creep-controlled failure is seen to be taken over by a crack growth failure in the high cycle region.

A combination of high frequency and high load can result in excessive dissipative heating in fiber reinforced thermoplastics to the extent that the failure is the result of thermal softening rather than fatigue [156,157]. Figure 1.4-58 shows temperature rise and apparent stress in two types of glass reinforced polypropylene subjected to a displacement-controlled flexural fatigue. The temperature rise is seen to be accompanied by a decrease of apparent stress, i.e., modulus reduction.

Fatigue life may be improved if the temperature increase is prevented by cooling. Compared with non-isothermal fatigue, isothermal fatigue failure is brittle and accompanied by more fiber fracture and less fiber pull-out. Also, the life in isothermal fatigue is longer, especially at high stresses, and shows less dependence on frequency (Figure 1.4-59) [156].

Better fiber-matrix bond and longer fiber length reduce the temperature increase and result in longer fatigue life. Thermoplastics with improved bonding show less fiber pull-out and tend to fail abruptly with less softening [156]. Such observation indicates that the heat generation is not only from the viscoelastic deformation of matrix but also from the friction between fibers and matrix.

The effect of moisture conditioning on room temperature fatigue properties of glass reinforced polybutyleneterephthalate (GRPBT) was studied by Di-Benedetto [154]. Specimens were preconditioned in three different environments: 100% RH for 100 hours (RH 100), boiling water for 192 hours (BW 192), and 70°C water for 504 hours (HW 504). The results on tensile properties are listed in Table 1.4-6. The BW 192 environment is seen to be most deleterious followed by the HW 504 environment. Although the RH 100 environment is responsible for a slight reduction in modulus and UTS, it improves failure strain. While a combination of elevated temperature and water leads to the embrittlement of the polymer, the smaller amount of water at room temperature simply plasticizes the polymer, increasing its ductility.

The failure strain in tension, ϵ_b, is plotted versus the number of cycles to failure in Figure 1.4-60 [136]. In the figure the first number next to a line denotes the type of environment (defined in Table 1.4-6) and the second number is the ratio of fatigue stress to UTS. The BW

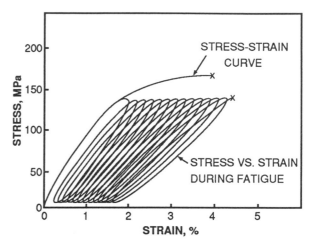

FIGURE 1.4-55. Stress-strain curves in fatigue and in static tension.

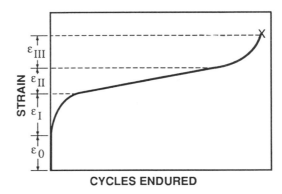

FIGURE 1.4-56. A schematic view of cyclic creep strain in fatigue.

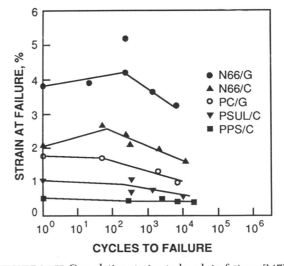

FIGURE 1.4-57. Cumulative strains to break in fatigue [147].

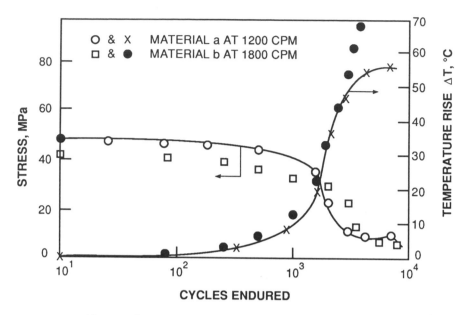

FIGURE 1.4-58. Changes of temperature and apparent stress in flexural fatigue of glass fiber reinforced polypropylene, $f = 20$ Hz [156].

192 environment is seen to substantially degrade the fatigue performance. Specimens preconditioned in BW 192 and HW 504 environments all failed in a brittle manner. On the other hand, the RH 100 environment improves fatigue life relative to specimens subject to no preconditioning (Environments 1 and 2 in Table 1.4-6). Recall that the schematic representation of cyclic creep strain shown in Figure 1.4-56 can be divided into three regions.

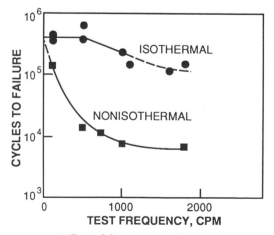

FIGURE 1.4-59. Effect of frequency on flexural fatigue life of glass fiber reinforced polypropylene [156].

The moisture-induced embrittlement of the polymer changes the initial softening strain ϵ_I and the steady state creep strain ϵ_{II}. For GRPBT under consideration ϵ_{III} was negligible [156] (see Table 1.4-7).

For specimens of high ductility, the total creep strain $(\epsilon_I + \epsilon_{II})$ is 50–70% of the tensile failure strain. The percentage decreases to 30–50% as the ductility diminishes. The matrix embrittlement reduces not only the tensile failure strain but also the cyclic creep strain.

S-N relations of various SMC composites have been summarized by Heimbuch and Sanders [158]. These relations are in general agreement with the trend shown in Figure 1.4-54. Composites with high fiber content such as SMC-R65 exhibit a steeper slope in the low cycle region while the slope does not change for composites with low fiber content. The slope of S-N relations is not changed much by temperature, indicating that the reduction of fatigue strength induced by temperature is in proportion to that of static strength.

Contrary to the behavior of injection molded thermoplastics such as glass reinforced polybutylene-tere-phthalate [154], the cyclic creep strain in SMC-R65 increases linearly with logarithmic fatigue cycles rather than fatigue cycles [158]. The cyclic creep strain at failure is in the order of 0.2% at 60% UTS; however, it is only half as much at 30% UTS.

Rubber toughening of matrix resin in SMC improves the resistance to matrix cracking and increases ultimate tensile strength if fiber volume fraction is low [159,160]. However, its effect on fatigue performance is mixed at present. Rowe and McGarry [159] studied the effect of matrix toughening by a liquid rubber [HYCAR Reactive Liquid Polymer (RLP)] using three different polyester resins in SMC-R30. The results in Table 1.4-8 clearly indicate strength improvements ranging from 22% to 82%. However, the B/UTS ratios for rubber toughened SMCs are all higher than for control formulations. It should be noted that, although the B/UTS ratios are higher, the fatigue strength at 10⁶ cycles is fairly independent of rubber toughening. The last entry in Table 1.4-8 is from reference [145]. Contrary to the first three SMC-R50 composites, the SMC-R25 shows an improvement in UTS without any change in B/UTS ratio.

Effect of Damage on Mechanical Properties

As in continuous fiber composites, fatigue damage in discontinuous fiber composites reduces both modulus and strength [150].

Owen and Howe [143] characterized the residual strength of chopped strand mat polyester composites subjected to tensile and fatigue loading. Damage accumulation was quantified in terms of fiber debonding and resin cracking. It was concluded that the development of resin cracking was primarily responsible for the progressive loss of tensile strength measured.

Typical modulus changes in fatigue of SMC-R65 are shown in Figure 1.4-61. The data for SMC-R50 by Denton [161] and Wang et al. [150] indicate that modulus reduction is fairly linear with logarithmic fatigue cycles with more reduction occurring at higher stresses.

The global nature of the damage in composites requires that attention be directed toward changes of global properties such as modulus and creep. Where design is dictated by stiffness, full information should be available on modulus as well as on strength.

1.4.5 Life Prediction and Data Analysis

The first ply failure of laminates is usually estimated by using the laminated plate theory in conjunction with an appropriate failure criterion. Some of the frequently used failure criteria are given in Equations (1.4-6) through (1.4-9).

Table 1.4-6. Effect of moisture conditioning on tensile properties of GRPBT [154].

Environment	UTS (MPa)	Failure Strain (%)	Modulus (GPa)	Toughnessᵇ (MPa)
1-RH 55ᵃ	121	3.2	7.72	2.68
2-RH 55ᵃ	121	2.9	7.24	2.46
3-RH 100	112	3.3	6.69	2.62
4-BW 192	77	1.5	6.48	0.69
5-HW 504	97	2.0	6.83	1.15

ᵃTwo different populations without any preconditioning.
ᵇArea under the stress-strain curve to failure.

Table 1.4-7. Effect of moisture embrittlement of GRPBT on cyclic creep strains [154]. Defined in Figure 1.4-56.

	Static Failure Strain ϵ_b (%)		
	$\epsilon_b > 3.3$	$3.1 > \epsilon_b > 2$	$1.8 > \epsilon_b > 1.3$
Cyclic Softening Strain ϵ_I (%)	1.0–1.6	0.4–0.9	0.2–0.5
Steady Creep Strain ϵ_{II} (%)	0.9–1.1	0.6–1.4	0.2–0.3
Total Cyclic Creep Strain $\epsilon_I + \epsilon_{II}$ (%)	1.9–2.7	1.0–2.3	0.4–0.8
$(\epsilon_I + \epsilon_{II})/\epsilon_b$	0.5–0.7	0.4–0.6	0.3–0.5

Table 1.4-8. Effect of rubber toughening on fatigue behavior of SMC [145,146].

Resin	Glass (wt%)	RLP (phr)	UTS (MPa)	B (MPa)	B/UTS
P-340/LP 40A	30	0	83	9.7	0.12
		8	101	13.9	0.14
P-340/P-701	30	0	53	4.2	0.08
		8	96	10.8	0.11
GR 13031/LP 40A	30	0	70	57	0.08
		8	91	9.1	0.10
PPG 50271/LP 40A	25	0	70	7.0	0.10
		4	83	8.3	0.10

P-340 (Rohm and Haas)
GR 13031 (USS Chemical)
PPG 50271 (PPG)
LP 40A (Union Carbide)
P-701 (Rohm and Haas)
RLP (BFGoodrich)

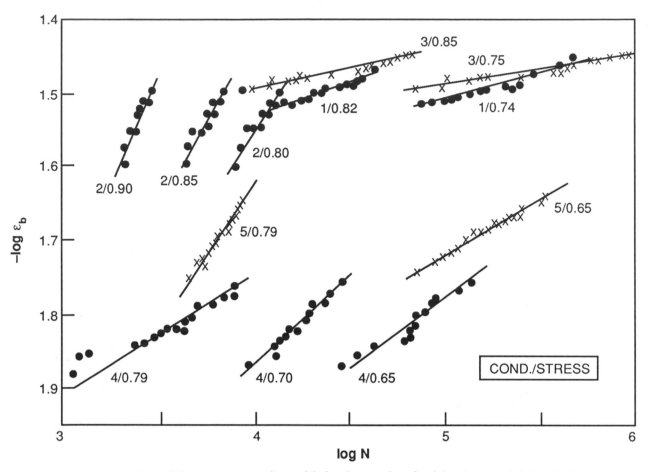

FIGURE 1.4-60. Static failure strain versus fatigue life for glass reinforced polybutyleneterephthalate [154].

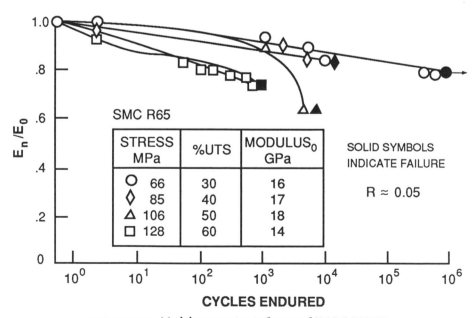

FIGURE 1.4-61. Modulus retention in fatigue of SMC-R65 [158].

The prediction of final failure is more difficult because the load redistribution resulting from ply failures is not exactly known. Consequently, several approaches have been proposed to model the in situ behavior of failed plies [84,162–166]. An incremental laminate analysis is then performed until all plies fail.

In fatigue, one may attempt to predict the fatigue strengths of laminates in terms of those of unidirectional composites the same way one would the static strength. Such an approach is conceptually simple; however, its implementation requires the knowledge of the in situ fatigue behavior of constituent plies which is not available at present. One exception is angle-ply ($\pm\theta$) laminates under uniaxial fatigue. In this case the laminate failure closely coincides with the ply failure.

Therefore, in this subsection we shall describe a laminated plate analysis for fatigue strengths of angle-ply laminates under uniaxial tension fatigue, assuming no change of elastic properties. The same approach can be applied to the first ply failure of general laminates in fatigue.

Next, two statistical life prediction models are introduced: one based on the concept of failure potential and the other on strength degradation. These models are analogous to the crack growth models for fatigue failure of homogeneous materials. However, since the failure of composites is not the result of a single dominant crack growth, crack length cannot be used to describe the strength-controlling state of composites. Therefore, failure potential and residual strength are chosen as two substitute parameters.

The statistical life prediction models provide a convenient means of analyzing fatigue data which invariably involve large scatter. An example will be given to illustrate the analysis procedure involved. Other direct methods of characterizing S-N relations are also discussed.

Angle-Ply Laminates

The strength of angle-ply laminates subjected to an axial tension can be predicted by using an appropriate failure criterion for the constituent plies and the relation between the applied stress and ply stresses. Denoting by σ_θ the stress applied in the 0° direction, the ply stresses in $[\pm\theta]_s$ angle-ply laminate are given by:

$$\sigma_L = k_L \sigma_\theta, \quad \sigma_T = k_T \sigma_\theta, \quad \sigma_{LT} = k_{LT} \sigma_\theta \quad (1.4\text{-}20)$$

Here the transfer coefficients are determined from the laminated plate theory. Specifically, they are given by [165]:

$$k_L = \frac{1}{2}\left[1 + \sec 2\theta - \frac{(u_1 + \sec 2\theta)\tan^2 2\theta}{u_2 + \tan^2 2\theta} \right] \quad (1.4\text{-}21)$$

$$k_T = \frac{1}{2}\left[1 - \sec 2\theta + \frac{(u_1 + \sec 2\theta)\tan^2 2\theta}{u_2 + \tan^2 2\theta} \right] \quad (1.4\text{-}22)$$

$$k_{LT} = -\frac{1}{2}\frac{(u_1 + \sec 2\theta)\tan 2\theta}{u_2 + \tan^2 2\theta} \quad (1.4\text{-}23)$$

$$u_1 = \frac{1 - E_L/E_T}{1 + 2\nu_{LT} + E_L/E_T} \quad (1.4\text{-}24)$$

$$u_2 = \frac{E_L/G_{LT}}{1 + 2\nu_{LT} + E_L/E_T} \quad (1.4\text{-}25)$$

The ply stresses are then substituted into a failure criterion and the resulting equation is solved for σ_θ.

Kim [167] investigated the applicability of the tensor polynomial to the failure of angle-ply graphite/epoxy laminates subjected to tension and compression. Figure 1.4-62 shows the ply stresses at failure normalized with respect to the corresponding uniaxial strengths. The curves are the predictions based on the tensor polynomial:

$$F_i\sigma_i + F_{ij}\sigma_i\sigma_j = 1 \quad (1.4\text{-}9)$$

with

$$F_{12} = -\frac{1}{2}\sqrt{F_{11}F_{22}} \quad (1.4\text{-}26)$$

It is seen that the ply stresses at failure show more interaction than predicted by the maximum stress criterion.

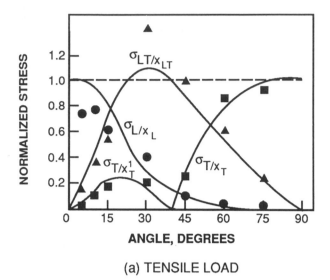

(a) TENSILE LOAD

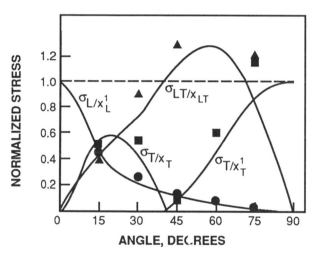

(b) COMPRESSIVE LOAD

FIGURE 1.4-62. Normalized ply stresses in angle-ply laminates at failure [167].

In fatigue, the problem is more complicated because the in situ behavior of constituent plies may change with cyclic loading. However, one can use Equations (1.4-20–1.4-25) as a first-order approximation. Once the ply stresses are known in terms of the laminate stress, a failure criterion such as Equation (1.4-9) can be used to predict fatigue strength of an angle-ply laminate from the uniaxial fatigue strengths of constituent plies. Note

that F_i and F_{ij} are related to the uniaxial strengths at N cycles by:

$$F_1 = \frac{1}{S_L} - \frac{1}{S_L'} , \quad F_{11} = \frac{1}{S_L S_L'}$$

$$F_2 = \frac{1}{S_T} - \frac{1}{S_T'} , \quad F_{22} = \frac{1}{S_T S_T'} \qquad (1.4\text{-}27)$$

$$F_{66} = \frac{1}{S_s^2}$$

where the prime indicates compression. If the stress ratio is R, the uniaxial fatigue strengths in tension and those in compression must be obtained at the same R.

The tensor polynomial criterion has not been used so far perhaps because of the difficulty of obtaining all the uniaxial fatigue strengths. Rotem and Hashin [165,168] used the Tsai-Hill criterion [Equation (1.4-8)] for various angle-ply glass/epoxy laminates. The prediction of fatigue strengths was as good as that of static strength.

Statistical Life Prediction: Failure Potential Model

As fatigue failure of composite laminates is the result of numerous types of damages, the life prediction methods based on the initiation and growth of a single dominant crack are not applicable. Therefore, we introduce a phenomenological parameter called the material age as a measure of the damage state. An analogy is found in the transition of the dislocation theory to the phenomenological plasticity theory in that plastic deformation is the result of the motion of numerous dislocations. The effect of individual dislocations on subsequent plastic deformation cannot at present be exactly described. Therefore, a macroscopic parameter called strain hardening parameter is introduced to phenomenologically describe the current state of plastically deformed material. It is in this analogy that we introduce the material age τ as a measure of the material degradation.

A failure process is often described by the failure rate that physically represents the probability of failure within a unit time interval [169]. If the probability of survival at material age τ is $R(\tau)$, the failure rate λ is related to R by:

$$\lambda = -\frac{dR/d\tau}{R} \qquad (1.4\text{-}28)$$

Note that material age τ has been used in place of real time t.

A typical variation of λ with τ is shown in Figure 1.4-63. The decreasing failure rate in the initial break-in period indicates failure being dominated by initial defects whereas the increasing failure rate in the final period is a sign of a wear-out process. A constant failure rate in the middle period results from a random failure process.

If λ is a known function of τ, the corresponding probability of survival is obtained from Equation (1.4-28) as:

$$R = \exp\left(-\int\lambda d\tau\right) \qquad (1.4\text{-}29)$$

Introducing a failure potential $\psi(\tau)$ such that

$$\lambda = d\psi/d\tau \qquad (1.4\text{-}30)$$

we can rewrite Equation (1.4-30) as:

$$R(\tau) = \exp\left[-\psi(\tau)\right] \qquad (1.4\text{-}31)$$

The relation between ψ and τ is an intrinsic material property and the effect of load history is through τ. In particular, we assume that material age τ under a load history $\sigma(\xi)$ is given by:

$$\tau = \int_0^t K[\sigma(\xi)]d\xi \qquad (1.4\text{-}32)$$

where K is called the aging rule. K is similar to the breakdown rule introduced by Coleman [169].

In the original formulation of the theory [169], the breakdown rule was independent of the type of load history. In the present development, however, the original assumption is relaxed so that the aging rule remains the same only within each type of load history. Thus, we choose:

$$K(\sigma) = c_1\gamma\sigma^{\gamma-1}, \ \gamma \geq 1 \qquad (1.4\text{-}33)$$

for tensile loading. An aging rule for fatigue loading will be discussed later.

As for the failure potential we choose a power law:

$$\psi(\tau) = c_2\tau^\alpha \qquad (1.4\text{-}34)$$

The corresponding failure rate is then:

$$\lambda = c_2\alpha\tau^{\alpha-1} \qquad (1.4\text{-}35)$$

Thus, a wear-out failure process is described by $\alpha > 1$ while a random failure process follows from $\alpha = 1$. The value of α less than unity points to a failure process controlled by initial defects.

Now consider a uniaxial tension test such that:

$$\sigma(t) = c_3t, \ t \geq 0 \qquad (1.4\text{-}36)$$

The resulting material age is:

$$\tau = c_1c_3^{\gamma-1}t^\gamma \qquad (1.4\text{-}37)$$

Therefore, the probability of survival at t is:

$$R(t) = \exp\left[-c_2\left(\frac{c_1}{c_3}\right)^\alpha(c_3t)^{\alpha\gamma}\right] \qquad (1.4\text{-}38)$$

If X is the stress at t, Equation (1.4-38) also represents the probability of survival at X. Furthermore,

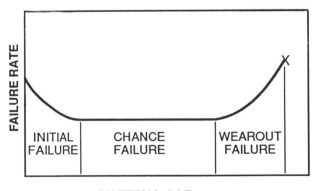

FIGURE 1.4-63. A schematic change of failure rate with material age τ.

noting that the cumulative distribution F is $1 - R$, we obtain the cumulative strength distribution $F_s(X)$ as:

$$F_s(X) = 1 - \exp\left[-\left(\frac{X}{X_0}\right)^{\alpha_s}\right] \quad (1.4\text{-}39)$$

where

$$\alpha_s = \alpha\gamma \quad (1.4\text{-}40)$$

and the characteristic strength X_0 is defined by:

$$X_0 = \left(\frac{c_3}{c_1}\right)^{1/\gamma} \cdot \left(\frac{1}{c_2}\right)^{1/\alpha_s} \quad (1.4\text{-}41)$$

To distinguish as much as possible between monotonic tension and fatigue behavior, we assume that fatigue loading is preceded by a monotonic tension to the maximum fatigue stress S. Then, in terms of the material age in fatigue, L, the life distribution $F_f(L)$ is given by:

$$F_f(L) = 1 - \exp\left\{-\left[\left(\frac{S}{X_0}\right)^{\gamma} + L\right]^{\alpha}\right\} \quad (1.4\text{-}42)$$

Equation (1.4-42) is a three-parameter Weibull distribution with a negative location parameter. $F_f(0)$ simply represents the probability of failure during the initial tension to S. The relation between L and fatigue cycles requires another aging rule. This will be discussed in the following subsection.

Statistical Life Prediction: Strength Degradation Model

Since composite failure is characterized by a multitude of cracks rather than a single dominant crack growth, the criticality of damage cannot be assessed by means of a single crack length. Therefore, we use residual strength as a measure of damage. The change of residual strength is postulated in analogy to the crack growth laws for homogeneous materials [132,136].

Specifically, the change of residual strength, X_r, is assumed to follow:

$$\frac{d\overline{X}_r}{dL} = -\frac{1}{\gamma}\overline{X}_r^{-\gamma+1} \quad (1.4\text{-}43)$$

where $\overline{X}_r = X_r/X_0$. Upon integration Equation (1.4-43) yields a relation between initial strength and residual strength:

$$\overline{X}_r^{\gamma} = \overline{X}^{\gamma} - L \quad (1.4\text{-}44)$$

where $\overline{X} = X/X_0$.

If a crack growth law is deterministic, the crack length at any time can be related to the initial crack length that in turn determines an initial strength. Such a damage process is called similar. Thus, one can establish a one-to-one relationship between initial strength and residual strength or lifetime under a similar fatigue damage process [170]. Although the strength of a composite cannot be related to a crack, we still assume that the strength degradation is similar. Since this assumption leads to a strength-life relation such that a statically strong specimen is also strong in fatigue, it is also called the equal rank assumption [171].

Now, suppose the initial strength distribution is given by [cf. Equation (1.4-39)]:

$$F_s(\overline{X}) = 1 - \exp\left(-\overline{X}^{\alpha_s}\right) \quad (1.4\text{-}45)$$

The residual strength distribution $F_r(\overline{X}_r)$ is then obtained by substituting Equation (1.4-44) into Equation (1.4-45):

$$F_r(\overline{X}_r) = 1 - \exp\left[-(\overline{X}_r^{\gamma} + L)^{\alpha_s/\gamma}\right] \quad (1.4\text{-}46)$$

Since fatigue failure occurs when residual strength reduces to S, the life distribution $F_f(L)$ follows upon substitution of $\overline{S}$ for $\overline{X}_r$ in Equation (1.4-46):

$$F_f(L) = 1 - \exp\left[-(\overline{S}^{\lambda} + L)^{\alpha_s/\gamma}\right] \quad (1.4\text{-}47)$$

where $\overline{S} = S/X_0$.

Equation (1.4-47) is exactly the same as Equation (1.4-43) if Equation (1.4-40) is used.

However, it should be recalled that the static strength distribution follows from the failure potential and the aging rule in the failure potential model whereas it is assumed a priori in the strength degradation model.

The life distribution in number of cycles to failure depends on the aging rule relating material age to number of cycles applied. Suppose the material age at failure, L, is related to the number of cycles to failure, N, by:

$$L = c_4\overline{S}^{\beta}N \quad (1.4\text{-}48)$$

The resulting life distribution is then:

$$F_f(N) = 1 - \exp\left[-\left(\bar{S}^\gamma + \frac{N}{N_0}\right)^{\alpha_s/\gamma}\right] \quad (1.4\text{-}49)$$

where

$$N_0 \bar{S}^\beta = \frac{1}{c_4} \quad (1.4\text{-}50)$$

On the other hand, if L is given by:

$$L = c_5 N \exp\left[\frac{\bar{S}}{B}\right] \quad (1.4\text{-}51)$$

then N_0 in Equation (1.4-49) is replaced by:

$$N_0 = \frac{1}{c_5} \exp\left[-\frac{\bar{S}}{B}\right] \quad (1.4\text{-}52)$$

Equation (1.4-52) can be put into a more familiar form:

$$\bar{S} = c_6 - \frac{B}{\log e} \log N_0 \quad (1.4\text{-}53)$$

where

$$c_6 = -B \frac{\log c_5}{\log e} \quad (1.4\text{-}54)$$

If the fatigue stress is lower than the characteristic strength, the first term inside the brackets in Equation (1.4-49) can be neglected and a two-parameter Weibull distribution is recovered. Equation (1.4-50) is none other than a power relation between the characteristic life N_0 and $\bar{S}$ while Equation (1.4-53) is an exponential relation.

The aging rules, Equations (1.4-48) and (1.4-51), are both independent of the load history and also do not change from specimen to specimen. Therefore, they may be called history-independent, uniform aging rules. A general aging rule may be formulated as follows.

Suppose material age L is related to number of cycles applied, n, by:

$$dL = D \zeta n^{\zeta-1} dn \quad (1.4\text{-}55)$$

Note that $\zeta > 1$ indicates an accelerating aging while $\zeta < 1$ represents a decelerating aging. A uniform aging is associated with $\zeta = 1$. The parameter D depends on loading parameters such as S, R, and f as well as on static strength X. Thus, a specimen with different static strength will age differently. Since finding such a general function is difficult, we rely on the static strength distribution, Equation (1.4-45), and a fatigue life distribution of the form:

$$F_f(N) = 1 - \exp\left[-\left(\bar{S}^{\alpha_s/\alpha} + \frac{N}{N_0}\right)^\alpha\right] \quad (1.4\text{-}56)$$

where $N_0 = N_0(S,R,f)$. Note that Equation (1.4-56) has been chosen in analogy to Equation (1.4-49). However, α is an independent variable not related to α_s or γ.

Noting from Equation (1.4-55) that

$$L = D n^\zeta \quad (1.4\text{-}57)$$

and using Equation (1.4-44), we can express the residual strength at n as:

$$\bar{X}_r^\gamma = \bar{X}^\gamma - D n^\zeta \quad (1.4\text{-}58)$$

At fatigue failure $\bar{X}_r$ and n respectively become:

$$\bar{X}_r = \bar{S}, \quad n = N \quad (1.4\text{-}59)$$

Therefore, D must satisfy:

$$D = \frac{\bar{X}^\gamma - \bar{S}^\gamma}{N^\zeta} \quad (1.4\text{-}60)$$

Because of the assumption of similar strength degradation, fatigue life N is related to static strength X by:

$$\bar{X}^{\alpha_s/\alpha} = \bar{S}^{\alpha_s/\alpha} + \frac{N}{N_0} \quad (1.4\text{-}61)$$

Combining Equations (1.4-60) and (1.4-61), we finally obtain:

$$D = \frac{1}{N_0^\zeta} \frac{\bar{X}^\gamma - \bar{S}^\gamma}{[\bar{X}^{\alpha_s/\alpha} - \bar{S}^{\alpha_s/\alpha}]^\zeta} \quad (1.4\text{-}62)$$

Table 1.4-9. Parameters assumed in fatigue models.

Reference	β	γ	ζ
[138]	–	–	–
[172]	–	–	1
[173]	–	α_s/α	1
[136]	$(\alpha_s/\alpha) + 2$	α_s/α	1
[174]	α_s/α	α_s/α	1

Substitution of Equation (1.4-62) into Equation (1.4-58) finally yields a relation between $\overline{X}_r$ and $\overline{X}$:

$$\overline{X}_r^\gamma = \overline{X}^\gamma - \frac{\overline{X}^\gamma - \overline{S}^\gamma}{[\overline{X}^{\alpha_s/\alpha} - \overline{S}^{\alpha_s/\alpha}]^\zeta}\left(\frac{n}{N_0}\right)^\zeta \qquad (1.4\text{-}63)$$

Equation (1.4-63) was derived by Yang and Jones [138] in a different manner. This equation can now be used in conjunction with the static strength distribution, Equation (1.4-45), to determine the residual strength distribution after n cycles by using the condition $F_r(\overline{X}_r) = F_s(\overline{X})$.

A complete description of D requires the knowledge of the dependence of N_0 on loading parameters. Yang and Jones [138] suggested the use of the normalized stress range $\Delta\overline{S}$ such that:

$$N_0(\Delta\overline{S})^\beta = \frac{1}{b} \qquad (1.4\text{-}64)$$

Note that Equation (1.4-64) is similar to Equation (1.4-50), and

$$\Delta\overline{S} = \overline{S}(1 - R) \qquad (1.4\text{-}65)$$

The general model can be reduced to simpler models by assigning a priori appropriate values to the parameters. If the material aging is independent of history, ζ should be unity. The resulting model was used by Yang and Jones [172] to study the effect of load sequence in cross-ply glass/epoxy laminates.

The model with $\zeta = 1$ and $\gamma = \alpha_s/\alpha$, as given in Equations (1.4-43) and (1.4-47), has been used by Hahn and Kim [132] and by Yang [173]. In this model the rate of material aging is independent of the static strength.

If the strength degradation is modeled after a crack growth law involving a stress intensity factor, β is no longer independent but should be equal to $(\alpha_s/\alpha) + 2$ [136]. This is known as the wear-out model and formed the basis of further development of the other models.

Sendeckyj [174] proposed to use $\beta = \alpha_s/\alpha$ in addition to $\zeta = 1$ and $\gamma = \alpha_s/\alpha$. All models, regardless of the level of simplification, fit the data well, indicating that data fitting alone is not discriminating enough. The assumed model parameters are summarized in Table 1.4-9.

Data Analysis: Strength Degradation Model

Fatigue life and residual strength data can be conveniently analyzed by using the models developed in the preceding subsections. Here we shall describe in detail how to determine the parameters α_s, α, γ, β, and b.

Suppose the following sets of data were obtained from a test program:

Static strengths : $\{X_i; i = 1,2, \ldots, m\}$

Lifetimes : $\{N_i, S_i, \Delta S_i; i = 1,2, \ldots, n\}$

Residual strengths : $\{X_{ri}, n_i, S_i, \Delta S_i; i = 1,2, \ldots, r\}$

The parameters α_s and X_0 are determined from the static strengths using Weibull statistics [181] based on the method of maximum likelihood described in the Appendix. Once X_0 is known, all stresses and strengths are normalized with respect to X_0. Next, the first three central moments are computed as:

$$m_1 = \frac{1}{m}\sum_{i=1}^m \overline{X}_i, \quad m_2 = \frac{1}{m}\sum_{i=1}^m (\overline{X}_i - m_1)^2 ,$$
$$(1.4\text{-}66)$$
$$m_3 = \frac{1}{m}\sum_{i=1}^m (\overline{X}_i - m_1)^3$$

Fatigue lifetimes are then converted into equivalent static strengths $\overline{X}_i'$ by using Equations (1.4-61) and (1.4-64):

$$\overline{X}_i'^{\alpha_s/\alpha} = \overline{S}_i^{\alpha_s/\alpha} + b\,\Delta\overline{S}_i^\beta N_i \qquad (1.4\text{-}67)$$

The corresponding moments are calculated as:

$$m_i' = \frac{1}{n} \sum_{i=1}^{n} \bar{X}_i', \quad m_2' = \frac{1}{n} \sum_{i=1}^{n} (\bar{X}_i' - m_1')^2 ,$$

$$(1.4\text{-}68)$$

$$m_3' = \frac{1}{n} \sum_{i=1}^{n} (\bar{X}_i' - m_1')^3$$

The parameters α, β and b are now determined by minimizing the mean square difference of the first three central moments:

$$\Delta = (m_1 - m_1')^2 + g_1(\sqrt{m_2} - \sqrt{m_2'})^2$$

$$+ g_2(\sqrt[3]{m_3} - \sqrt[3]{m_3'})^2 \qquad (1.4\text{-}69)$$

Here the relative importance of matching the mean, the standard deviation, and the skewness is varied by assigning appropriate weighting values g_1 and g_2.

The determination of ζ and γ is algebraically more involved. First, the equivalent static strengths $\bar{X}_i''$ corresponding to $\bar{X}_{ri}$ should be obtained from Equation (1.4-63) through iteration:

$$\bar{X}_{ri}^{\gamma} = \bar{X}_i''^{\gamma} - \frac{\bar{X}_i''^{\gamma} - \bar{S}_i^{\gamma}}{[\bar{X}_i''^{\alpha_s/\alpha} - \bar{S}_i^{\alpha_s/\alpha}]^{\zeta}} (b \Delta \bar{S}_i^{\beta} n_i)^{\zeta} \, (1.4\text{-}70)$$

The first three moments are then:

$$m_1'' = \frac{1}{r} \sum_{i=1}^{r} \bar{X}_i'', \quad m_2'' = \frac{1}{r} \sum_{i=1}^{r} (\bar{X}_i'' - m_1'')^2 ,$$

$$(1.4\text{-}71)$$

$$m_3'' = \frac{1}{r} \sum_{i=1}^{r} (\bar{X}_i'' - m_1'')^3$$

Again, the parameters γ and ζ are chosen such that the following mean square difference is a minimum:

$$\Delta = (m_1 - m_1'')^2 + g_1(\sqrt{m_2} - \sqrt{m_2''})^2$$

$$+ g_2(\sqrt[3]{m_3} - \sqrt[3]{m_3''})^2 \qquad (1.4\text{-}72)$$

In the special case where $\zeta = 1$ and $\gamma = \alpha_s/\alpha$, Equation (1.4-70) reduces to:

$$\bar{X}_i'^{\alpha_s/\alpha} = \bar{X}_{ri}^{\alpha_s/\alpha} + b \Delta \bar{S}_i^{\beta} n_i \qquad (1.4\text{-}73)$$

The equivalent strengths from Equation (1.4-73) are then combined with those from Equation (1.4-67), and Equation (1.4-68) is replaced by:

$$m_1' = \frac{1}{n+r} \sum_{i=1}^{n+r} X_i', \quad m_2' = \frac{1}{n+r} \sum_{i=1}^{n+r} (\bar{X}_i' - m_1')^2 ,$$

$$(1.4\text{-}74)$$

$$m_3' = \frac{1}{n+r} \sum_{i=1}^{n+r} (\bar{X}_i' - m_1')^3$$

The determination of α, β and b is based on Equation (1.4-69). Example [173]:

For [0/45/90/$-$45/90/45/0] T300/934 Gr/Ep laminate, static strengths and results of fatigue scan are listed in Tables 1.4-10 and 1.4-11, respectively. Furthermore, we assume the strength degradation to be history-independent and uniform so that $\zeta = 1$, $\gamma = \alpha_s/\alpha$.

The strength shape parameter and characteristic strength are determined from the strength data in Table 1.4-10 by the method of maximum likelihood as

$$\alpha_s = 24.95, \quad X_0 = 70.7 \text{ ksi} \qquad (1.4\text{-}75)$$

Using Equations (1.4-66) through (1.4-69) with $g_1 = g_2 = 1$, we obtain

$$\alpha_s/\alpha = 12, \quad \beta = 12.27, \quad b = 2.703 \times 10^{-4}$$

$$(1.4\text{-}76)$$

The cumulative distribution for the static strength $\bar{X}$ and the equivalent strength $\bar{X}'$ are shown in Figure

Table 1.4-10. Static strengths, ksi.

73.0	69.3	69.3	71.8	74.2	75.4	67.6
64.4	66.0	72.0	72.6	70.6	62.0	71.4
70.6	69.9	64.6	69.9	65.2	69.9	69.7
68.0	64.6	66.0	7.13			

$m = 25$.

Table 1.4-11. Fatigue scan data.

Maximum Stress S (ksi)	62	62	62	58	58	58	54	54
Stress Range ΔS (ksi)	72	72	72	68	68	68	64	64
Cycles to Failure, N	810	1127	10	4840	4980	1675	10500	11055
Maximum Stress S (ksi)	54	50	50	50	46	46	46	42
Stress Range ΔS (ksi)	64	60	60	60	56	56	56	52
Cycles to Failure, N	6997	10651	16030	10000	25520	36500	78710	130720
Maximum Stress S (ksi)	42	42	38	38	38	34	34	34
Stress Range ΔS (ksi)	52	52	48	48	48	44	44	44
Cycles to Failure, N	78290	188887	111600	414560	485190	1047000	1322440	2104510
Maximum Stress S (ksi)	39	58	58	50	50	50	38	38
Stress Range ΔS (ksi)	74	74	74	66	66	66	54	54
Cycles to Failure, N	1402	3251	1010	10906	11445	3981	213539	55380
Maximum Stress S (ksi)	38	34	34	34				
Stress Range ΔS (ksi)	54	50	50	50				
Cycles to Failure, N	43485	123672	749444	638880				

1.4-64. The cumulative distribution of the equivalent strength is seen to be close to that of the static strength.

For a fixed value of minimum fatigue stress S_{min} Equation (1.4-6l) describes a relation between ΔS and N:

$$N = \frac{1}{b\,\Delta\overline{S}^\beta}\,[\overline{X}^{\alpha_s/\alpha} - (\Delta\overline{S} + \overline{S}_{min})^{\alpha_s/\alpha}] \quad (1.4\text{-}77)$$

For the median strength $\overline{X} = 0.985$ for which $F_s(\overline{X}) = 0.5$, Equation (1.4-77) is compared with the data in Figure 1.4-65. The data are those in Table 1.4-11 with $S_{min} = -10$ ksi.

Since $\zeta = 1$ and $\gamma = \alpha_s/\alpha$ in the present case, the residual strength distribution after n cycles is given by:

$$F_r(\overline{X}_r) = F_s(\overline{X}) = 1 - \exp[-(\overline{X}_r^{\alpha_s/\alpha} + b\,\Delta\overline{S}^\beta n)^\alpha] \quad (1.4\text{-}78)$$

The distribution of strength of specimens surviving n cycles is then:

$$F_r(\overline{X}_r\,|\,\overline{S}) = 1 - \exp[-(\overline{X}_r^{\alpha_s/\alpha} + b\,\Delta\overline{S}^\beta n)^\alpha$$
$$+ (\overline{S}^{\alpha_s/\alpha} + b\,\Delta\overline{S}^\beta n)^\alpha] \quad (1.4\text{-}79)$$

The cumulative distribution $F_r(\overline{X}_r\,|\,\overline{S})$ is compared with two sets of residual strength data in Figure 1.4-66. Here one set is after 14,400 cycles at $\Delta S = 58$ ksi and $S = 42$ ksi, and the other after 2,150 cycles at $\Delta S = 66$ ksi and $S = 50$ ksi. Finally, the cumulative life distribution:

$$F_f(N) = 1 - \exp[-(\overline{S}^{\alpha_s/\alpha} + b\,\Delta\overline{S}^\beta N)^\alpha] \quad (1.4\text{-}80)$$

is compared with experimental data at various combinations of $\Delta\overline{S}$ and $\overline{S}$ in Figure 1.4-67. It is seen that

Equations (1.4-77), (1.4-79) and (1.4-80) with the parameters determined from the static strength data and the fatigue life scan data can reasonably well represent the *S-N* relations, the residual strength distributions, and the life distributions, respectively.

Data Analysis: S-N Characterization

In most cases fatigue tests are performed at sufficiently low stresses so that almost no specimens are expected to fail in a preloading to $\bar{S}$, i.e., $F_s(\bar{S}) \approx 0$. Therefore, the life distribution Equation (1.4-56) can be approximated by:

$$F_f(N) = 1 - \exp\left[- \left(\frac{N}{N_0} \right)^\alpha \right] \qquad (1.4\text{-}81)$$

where N_0 is defined by Equation (1.4-64):

$$N_0 \Delta \bar{S}^\beta = 1/b \qquad (1.4\text{-}82)$$

Thus, a complete characterization of *S-N* behavior is accomplished by determining the parameters α, β and b.

Let m be the number of stress ranges tested and the lifetime data at $\Delta \bar{S}_i$ be denoted by $\{N_{i1}, N_{i2}, \ldots, N_{ini}\}$. The number of run-out specimens at $N = R_i$ is denoted by r_i so that the total number of specimens tested at $\Delta \bar{S}_i$ is $n_i + r_i$.

The lifetime distribution at $\Delta \bar{S}_i$ is then obtained by using the method of maximum likelihood to determine α_i and N_{0i} (see Appendix). The least squares fit is then performed on the set $\{\Delta \bar{S}_i, N_{0i}; i = 1, 2, \ldots, m\}$ to obtain β and b:

$$\beta = - \frac{m\Sigma(\log \Delta \bar{S}_i)(\log N_{0i}) - \Sigma \Delta \bar{S}_i \Sigma \log N_{0i}}{m\Sigma(\log \Delta \bar{S}_i)^2 - (\Sigma \log \Delta \bar{S}_i)^2} \qquad (1.4\text{-}83)$$

$$\log b = - \frac{\Sigma(\log \Delta \bar{S}_i)^2 \Sigma \log N_{0i}}{m\Sigma(\log \Delta \bar{S}_i)^2 - (\Sigma \log \Delta \bar{S}_i)^2}$$
$$+ \frac{\Sigma \log \Delta \bar{S}_i \Sigma(\log \Delta \bar{S}_i)(\log N_{0i})}{m\Sigma(\log \Delta \bar{S}_i)^2 - (\Sigma \log \Delta \bar{S}_i)^2} \qquad (1.4\text{-}84)$$

where the summations are over i from 1 to m. Note that the shape parameter α is allowed to vary with the stress range in the foregoing method.

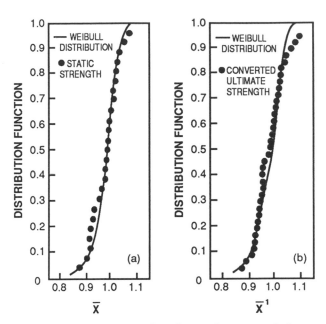

FIGURE 1.4-64. Static strength and equivalent strength distribution [173].

The significance of the variation of α with $\Delta \bar{S}$ can be investigated by using the two-sample test described by Thomas and Bain [175]. Let α_{max} and α_{min} be the maximum and minimum values obtained for α. It is further assumed that they are from two samples of equal size n.

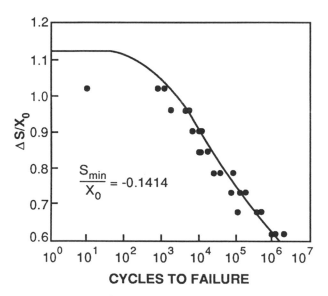

FIGURE 1.4-65. *S-N* relation for 50% probability of failure compared with data.

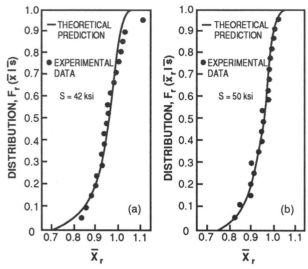

FIGURE 1.4-66. Conditional residual strength distribution, $S_{min} = -16$ ksi: (a) $S = 42$ ksi, $n = 14,400$ cycles; (b) $S = 50$ ksi, $n = 2,150$ cycles [173].

If both α_{max} and α_{min} are from the same population, then:

$$\frac{\alpha_{max}}{\alpha_{min}} < B(\gamma,n) \qquad (1.4\text{-}85)$$

for a given confidence level γ. Values of B are listed in Table 1.4-12 for a confidence level of 0.98 [175].

If α does not depend on $\Delta \overline{S}$, all data can be pooled to find a common shape parameter. Three methods of estimating the common shape parameter are described in the following [176,177].

The first method is to simply take the average of α_i, $i = 1,2,3, \ldots, m$:

$$\bar{\alpha} = \frac{1}{m} \sum_{i=1}^{m} \alpha_i \qquad (1.4\text{-}86)$$

The second method is to normalize the data at each $\Delta \overline{S}_i$ by N_{0i} and then pool the data. The corresponding maximum likelihood estimate is then obtained from:

$$\frac{1}{\tilde{\alpha}} = \frac{\displaystyle\sum_{i=1}^{m} \sum_{j=1}^{n_i} \overline{N}_{ij}^{\tilde{\alpha}} \, \ell n \, \overline{N}_{ij} + \sum_{i=1}^{m} r_i \overline{R}_i^{\tilde{\alpha}} \, \ell n \, \overline{R}_i}{\displaystyle\sum_{i=1}^{m} \sum_{j=1}^{n_i} \overline{N}_{ij}^{\tilde{\alpha}} + \sum_{i=1}^{m} r_i \overline{R}_i^{\tilde{\alpha}}}$$

$$- \frac{1}{n_0} \sum_{i=1}^{m} \sum_{j=1}^{n_i} \ell n \, \overline{N}_{ij} \qquad (1.4\text{-}87)$$

where

$$\overline{N}_{ij} = N_{ij}/N_{0i} \qquad (1.4\text{-}88)$$

$$\overline{R}_{ij} = R_i/N_{0i} \qquad (1.4\text{-}89)$$

$$n_0 = \sum_{i=1}^{m} n_i \qquad (1.4\text{-}90)$$

The third is called the joint maximum likelihood estimate $\hat{\alpha}$. It is determined from:

$$\frac{1}{\hat{\alpha}} = \frac{1}{m} \sum_{i=1}^{m} \left(\frac{\displaystyle\sum_{j=1}^{n_i} N_{ij}^{\hat{\alpha}} \, \ell n \, N_{ij} + r_i R_i^{\hat{\alpha}} \, \ell n \, R_i}{\displaystyle\sum_{j=1}^{n_i} N_{ij}^{\hat{\alpha}} + r_i R_i^{\hat{\alpha}}} \right.$$

$$\left. - \frac{1}{n_i} \sum_{j=1}^{n_i} \ell n \, N_{ij} \right) \qquad (1.4\text{-}91)$$

Once α is estimated from one of Equations (1.4-86), (1.4-87), and (1.4-91), the characteristic lifetimes are determined by:

$$N_{0i} = \left(\frac{1}{n_i} \sum_{j=1}^{n_i} N_{ij}^{\alpha} + r_i R_i^{\alpha} \right)^{1/\alpha} \qquad (1.4\text{-}92)$$

These values are then substituted into Equations (1.4-83) and (1.4-84) to obtain β and b.

Whitney [178] used Equation (1.4-87) to analyze more extensive data for the same composite as used in Tables 1.4-10 and 1.4-11. As for the characteristic lifetimes he did not use Equation (1.4-92); rather, he obtained the normalized characteristic lifetime $\overline{N}_0$ from the normalized data $\overline{N}_{ij}$. The characteristic lifetimes to be fit by Equation (1.4-82) were then taken equal to N_0 mul-

Table 1.4-12. Values of B (γ,n) for $\gamma = 0.98$.

Sample Size n	5	10	15	20	100
$B(\gamma,n)$	3.550	2.213	1.870	1.703	1.266

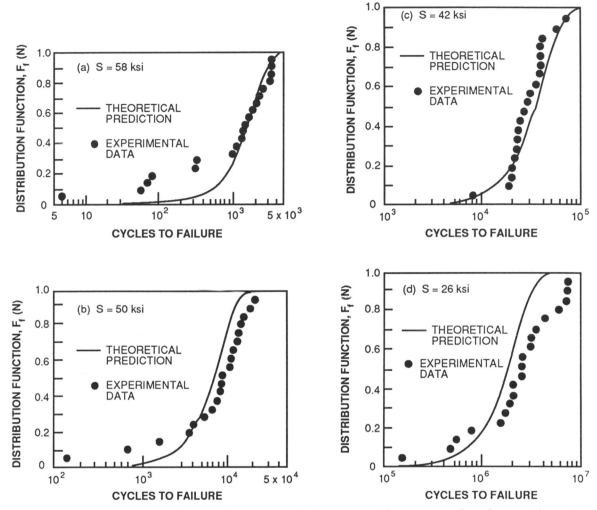

FIGURE 1.4-67. Life distributions, $S_{min} = -16$ ksi: (a) $S = 58$ ksi; (b) $S = 50$ ksi; (c) $S = 42$ ksi; (d) $S = 26$ ksi [173].

tiplied by N_{0i} associated with α_i, but not $\hat{\alpha}$. The parameters thus obtained are:

$$\alpha = 1.45, \quad \beta = 9.705, \quad b = 3.497 \times 10^{-3} \quad (1.4\text{-}93)$$

These parameters are quite different from those in Equations (1.4-76). It is to be noted that the values in Equation (1.4-93) are from a much wider data base than those in Equations (1.4-76).

If only a limited number of lifetimes is available at each stress range, then an S-N relation of the form:

$$N \, \Delta \bar{S}^{\beta} = 1/b \quad (1.4\text{-}94)$$

may be determined first by the method of least squares. The lifetimes $\{N_{ij}: j = 1,2,3 \ldots ,n_i\}$ are then normal-

ized with respect to the lifetime N_i predicted from Equation (1.4-94):

$$\bar{N}_{ij} = N_{ij}/N_i \quad (1.4\text{-}95)$$

where

$$N_i = \frac{1}{b} \, \Delta \bar{S}_i^{-\beta} \quad (1.4\text{-}96)$$

The shape parameter α and the normalized characteristic lifetime $\bar{N}_0$ are obtained from the normalized data $\{\bar{N}_{ij}: i = 1,2, \ldots ,m; j = 1,2, \ldots ,n_i\}$. The characteristic lifetime at $\Delta \bar{S}_i$ is then given by:

$$N_{0i} = \bar{N}_0 N_i \quad (1.4\text{-}97)$$

This method was used by Hahn and Kim [50].

As is clear by now there are a variety of methods available to characterize *S-N* relations. Depending on the availability of data one method may be more practicable than others. In cases where any method can be used, it is not clear at present which method is the best.

Load Sequence Effects

The strength degradation model can be used to account for the effect of load history. One of the simple load histories of interest is a cyclic loading of one amplitude followed by a cyclic loading of another.

Specifically, consider a load history where n_1 cycles at $\Delta \overline{S}_1$ and $\overline{S}_1$ are followed by another cyclic loading at $\Delta \overline{S}_2$ and $\overline{S}_2$ until failure. The residual strength at the end of the first n_1 cycles is given by:

$$\overline{X}_r^\gamma(n_1) = \overline{X}^\gamma - D_1 n_1^\zeta \qquad (1.4\text{-}98)$$

After additional n_{12} cycles at $\Delta \overline{S}_2$ and $\overline{S}_2$, the residual strength becomes:

$$\overline{X}_r^\gamma(n_1 + n_{12}) = \overline{X}_r^\gamma(n_1) - D_2[(n_1 + n_{12})^\zeta - n_1^\zeta]$$

$$= \overline{X}^\gamma - D_1 n_1^\zeta - D_2[(n_1 + n_{12})^\zeta - n_1^\zeta] \qquad (1.4\text{-}99)$$

In Equations (1.4-98) and (1.4-99) D_1 and D_2 are given by:

$$D_1 = \frac{1}{N_{01}^\zeta} \frac{\overline{X}^\gamma - \overline{S}_1^\gamma}{[\overline{X}^{\alpha s/\alpha} - \overline{S}_1^{\alpha s/\alpha}]^\zeta} \qquad (1.4\text{-}100)$$

$$D_2 = \frac{1}{N_{02}^\zeta} \frac{\overline{X}^\gamma - \overline{S}_2^\gamma}{[\overline{X}^{\alpha s/\alpha} - \overline{S}_2^{\alpha s/\alpha}]^\zeta} \qquad (1.4\text{-}101)$$

The residual strength distribution can now be determined from Equation (1.4-99) and $F_r(\overline{X}_r) = F_s(\overline{X})$.

The number of cycles to failure N_{12} is determined from the condition:

$$\overline{X}_r(n_1 + N_{12}) = \overline{S}_2 \qquad (1.4\text{-}102)$$

Therefore,

$$N_{12} = \left\{ \frac{1}{D_2} [\overline{X}^\gamma - \overline{S}_2^\gamma - (D_1 - D_2)n_1^\zeta] \right\}^{1/\zeta} - n_1 \qquad (1.4\text{-}103)$$

The life distribution is then obtained by combining Equation (1.4-103) with $F_f(N_{12}) = F_s(\overline{X})$.

To study the effect of load sequence, we first apply n_2 cycles at $\Delta \overline{S}_2$ and then n_{21} cycles at $\Delta \overline{S}_1$. The resulting residual strength is:

$$\overline{X}_r^\gamma(n_2 + n_{21}) = \overline{X}^\gamma - D_2 n_2^\zeta - D_1[(n_2 + n_{21})^\zeta - n_2^\zeta] \qquad (1.4\text{-}104)$$

When the same number of cycles is applied at each stress range, the difference between residual strengths given by Equations (1.4-99) and (1.4-104) becomes:

$$\overline{X}_r^\gamma(n_1 + n_{12}) - \overline{X}_r^\gamma(n_2 + n_{21}) = 2(D_1 - D_2)(2^{\zeta-1} - 1)n^\zeta \qquad (1.4\text{-}105)$$

where

$$n = n_1 = n_2 = n_{12} = n_{21} \qquad (1.4\text{-}106)$$

Thus, if $\zeta > 1$, i.e., under accelerating aging, the low-high sequence ($\Delta \overline{S}_1 < \Delta \overline{S}_2$) is more damaging because $D_1 < D_2$. Otherwise, the reverse prevails. If $\zeta = 1$, load sequence has no effect on residual strength degradation [138], and the residual strength is:

$$\overline{X}_r^\gamma(n_1 + n_{12}) = \overline{X}^\gamma - D_1 n_1 - D_2 n_{12} \qquad (1.4\text{-}107)$$

The fatigue life N_{12} is linearly related to n_1,

$$N_{12} = \frac{\overline{X}^\gamma - \overline{S}_2^\gamma}{D_2} - \frac{D_1}{D_2} n_1$$

$$= N_{02}(\overline{X}^{\alpha s/\alpha} - \overline{S}_2^{\alpha s/\alpha})$$

$$- \frac{N_{02}}{N_{01}} \frac{\overline{X}^{\alpha s/\alpha} - \overline{S}_2^{\alpha s/\alpha}}{\overline{X}^{\alpha s/\alpha} - \overline{S}_1^{\alpha s/\alpha}} \frac{\overline{X}^\gamma - \overline{S}_1^\gamma}{\overline{X}^\gamma - \overline{S}_2^\gamma} n_1 \qquad (1.4\text{-}108)$$

Now consider the Miner-Palmgren's damage sum defined by:

$$Q = \frac{n_1}{N_1} + \frac{N_{12}}{N_2} \qquad (1.4\text{-}109)$$

Here N_1 and N_2 are the fatigue lives at $\Delta \overline{S}_1$ and $\Delta \overline{S}_2$, respectively,

$$N_1 = \frac{1}{D_1} (\overline{X}^\gamma - \overline{S}_1^\gamma) \qquad (1.4\text{-}110)$$

$$N_2 = \frac{1}{D_2} (\overline{X}^\gamma - \overline{S}_2^\gamma) \qquad (1.4\text{-}111)$$

Substituting Equations (1.4-110) and (1.4-111) into Equation (1.4-109) leads to:

$$Q = 1 + D_1 n_1 \left(\frac{1}{\overline{X}^\gamma - \overline{S}_1^\gamma} - \frac{1}{\overline{X}^\gamma - \overline{S}_2^\gamma} \right) \quad (1.4\text{-}112)$$

Since $\overline{X}$ is a statistical variable, so is the Miner-Palmgren's damage sum. Q becomes unity when $\overline{S}_1 = \overline{S}_2$, as expected. However, in the low-high load sequence, i.e., $\overline{S}_1 < \overline{S}_2$, Q is less than unity whereas in

the high-low load sequence, Q is greater than unity. That is,

$$Q < 1 \ \text{if} \ \overline{S}_1 < \overline{S}_2$$
$$Q = 1 \ \text{if} \ \overline{S}_1 = \overline{S}_2 \quad (1.4\text{-}113)$$
$$Q > 1 \ \text{if} \ \overline{S}_1 > \overline{S}_2$$

Also, the deviation of Q from unity increases with the first fatigue cycles n_1. It is interesting to note that, al-

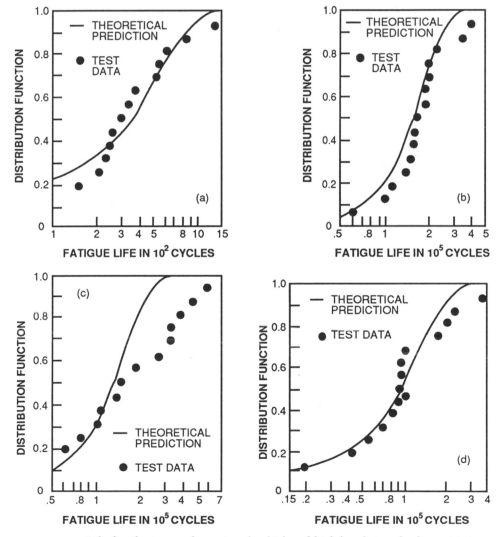

FIGURE 1.4-68. Life distributions under various low-high and high-low fatigue loadings: (a) $S_1 = 290$ MPa, $n_1 = 2000$, $S_2 = 386$ MPa; (b) $S_1 = 338$ MPa, $n_1 = 250$, $S_2 = 241$ MPa; (c) $S_1 = 386$ MPa, $n_1 = 250$, $S_2 = 241$ MPa; (d) $S_1 = 290$ MPa, $n_1 = 10,000$, $S_2 = 241$ MPa [172].

though the strength degradation is independent of load sequence, the Miner's damage sum does depend on it.

Yang and Jones [172] applied the model with $\zeta = 1$ to the fatigue data for a cross-ply E-Gl/Ep laminate obtained by Broutman and Sahu [93]. The parameters determined are:

$$\alpha_s = 15.6, \quad X_0 = 464 \text{ MPa}$$

$$\alpha = 2.6, \quad \beta = 11.41, \quad b = 1.618 \times 10^{-2}$$

(1.4-114)

The resulting life distributions $F_f(N_{12})$ are compared with the experimental data in Figure 1.4-68. A fairly good agreement is seen between the theory and the data for all combinations of low-high and high-low loadings considered.

1.4.6 Appendix: Weibull Distribution

Consider an ordered set of lifetime data $\{t_1 \leq t_2 \leq \cdots \leq t_r\}$ obtained from n specimens. The test was stopped after a failure at t_r. Thus the number of failed specimens is r, fewer than the total number n.

The probability $R(t_i)$ of survival at t_i, $1 \leq i \leq r$, experimentally is given by the media rank:

$$R(t_i) = 1 - \frac{i - 0.3}{n + 0.4}$$

On the other hand, the distribution of lifetimes analytically is given by a Weibull distribution of the form:

$$R(t) = \exp\left[-\left(\frac{t}{t_0} \right)^{\alpha} \right]$$

Here α and t_0 are called the shape parameter and characteristic lifetime, respectively. Note that $R(t)$ satisfies the limiting conditions $R(0) = 1$ and $R(\infty) = 0$.

The parameters α and t_0 can be determined by using the method of maximum likelihood, i.e., by maximizing the probability of the obtained data being realized [179]. The final equations are:

$$\frac{1}{\alpha} = \frac{\sum\limits_{i=1}^{r} t_i^{\alpha} \ln t_i + (n - r)t_r^{\alpha} \ln t_r}{\sum\limits_{i=1}^{r} t_i^{\alpha} + (n - r)t_r^{\alpha}} - \frac{1}{r} \sum\limits_{i=1}^{r} \ln t_i$$

$$t_0 = \left[\frac{1}{r} \sum\limits_{i=1}^{r} t_i^{\alpha} + (n - r)t_r^{\alpha} \right]^{1/\alpha}$$

The shape parameter provides valuable information as to the underlying failure processes. The failure rate, which represents the probability of failure per unit time interval, is given by:

$$\lambda = -\frac{dR/dt}{R} = \frac{1}{t_0} \left(\frac{t}{t_0} \right)^{\alpha-1}$$

Thus, as the shape parameter α is larger than, equal to, or smaller than unity, the failure rate increases, remains constant, or decreases with time. An increasing failure rate indicates a wear-out type of failure process while a decreasing failure rate can be taken as a sign of failure being dominated by initial defects. A constant failure rate results from a random failure process [180].

The mean $\bar{t}$ and coefficient of variation (C_V) of the Weibull distribution are given in terms of α and t_0 as follows:

$$\bar{t} = t_0 \Gamma\left(\frac{1}{\alpha} + 1 \right)$$

$$C_V = \left[\frac{\Gamma\left(\frac{2}{\alpha} + 1 \right)}{\Gamma^2\left(\frac{1}{\alpha} + 1 \right)} - 1 \right]^{1/2}$$

Here Γ is the gamma function. Note that C_V is independent of t_0. For α larger than unity, C_V can be approximated by

$$C_V \approx \frac{1.2}{\alpha}$$

Thus, α can readily be estimated from C_V.

1.4.7 References

1. BROEK, D. *Elementary Engineering Fracture Mechanics*. 4th revised ed., Martinus Nijhoff (1986).

2. YOKOBORI, T. *Strength, Fracture and Fatigue of Materials*. P. Noordhoff, Ltd., The Netherlands, p. 218 (1964).

3. REIFSNIDER, K. L. and R. Jamison. "Fracture of Fatigue-Loaded Composite Laminates," *Int. J. Fatigue*, 5:187 (1982).

4. MASTERS, J. E. and K. L. Reifsnider. "An Investigation of Cumulative Damage Development in Quasi-Isotropic Graphite/Epoxy Laminates," *Damage of Composite Materials*, ASTM STP 775, p. 40 (1982).

5. CARLSSON, L., C. Eidefeldt and T. Mohlin. "Influence of Sublaminate Cracks on the Tension Fatigue Behavior of a Graphite/Epoxy Laminate," *Composite Materials: Fatigue and Fracture*, ASTM STP 907, p. 361 (1986).

6. SALKIND, M. S. In *Composite Materials: Testing and Design (Second Conference)*, ASTM STP 497, American Society for Testing and Materials, p. 143 (1972).

7. OWEN, M. J. In *Composite Materials, Vol. 5, Fracture and Fatigue*. L. J. Broutman, ed., Academic Press, NY, p. 313 (1974).

8. OWEN, M. J. In *Composite Materials, Vol. 5, Fracture and Fatigue*. L. J. Broutman, ed., Academic Press, NY, p. 341 (1974).

9. OWEN, M. J. In *Composite Materials, Vol. 5, Fracture and Fatigue*. L. J. Broutman, ed., Academic Press, NY, p. 371 (1974).

10. HAHN, H. T. In *Composite Materials: Testing and Design (Fifth Conference)*, ASTM STP 674, S. W. Tsai, ed., American Society for Testing and Materials, p. 383 (1979).

11. STINCHCOMB, W. W. and K. L. Reifsnider. In *Fatigue Mechanisms*, ASTM STP 675, J. T. Fong, ed., American Society for Testing and Materials, p. 762 (1979).

12. METCALFE, A. G. and M. J. Klein. In *Composite Materials, Vol. 1, Interfaces in Metal Matrix Composites*. A. G. Metcalfe, ed., Academic Press, NY, p. 125 (1974).

13. COLLINS, B. R., W. D. Brentnall and I. J. Toth. In *Failure Modes in Composites*. I. J. Toth, ed., The Metallurgical Society of the American Institute of Mining, Metallurgical and Petroleum Engineers, p. 103 (1973).

14. STEELE, J. H. and H. W. Herring. In *Failure Modes in Composites*. The Metallurgical Society of the American Institute of Mining, Metallurgical and Petroleum Engineers, p. 343 (1973).

15. AWERBUCH, J. and H. T. Hahn. *Experimental Mechanics*, 20:334 (1980).

16. ROSEN, B. W. In *Fiber Composite Materials*. American Society for Metals, p. 37 (1965).

17. GUCER, D. E. and J. Gurland. *Journal of the Mechanics and Physics of Solids*, 10:365 (1962).

18. MULLIN, J., J. M. Berry and A. Gatti. *Journal of Composite Materials*, 2:82 (1968).

19. STREET, K. N. and J. P. Ferte. In *Proceedings of the 1975 International Conference on Composite Materials, Vol. 1*. The Metallurgical Society of the American Institute of Mining, Metallurgical and Petroleum Engineers, p. 137 (1976).

20. MILLS, G. J., G. Brown and D. Waterman. In *Proceedings of the 1975 International Conference on Composite Materials, Vol. 2*. The Metallurgical Society of the American Institute of Mining, Metallurgical and Petroleum Engineers, p. 222 (1976).

21. SCHUSTER, D. M. and E. Scala. In *Fundamental Aspects of Fiber Reinforced Plastic Composites*. R. T. Schwartz and H. S. Schwartz, eds., Interscience Pub., NY, p. 45 (1968).

22. SHOCKEY, P. D. and K. E. Hofer. Technical Report AFML-TR-70-108, Vol. II, Air Force Materials Laboratory (1972).

23. MENKE, G. D. and I. J. Toth. Technical Report AFML-TR-71-102, Air Force Materials Laboratory (1971).

24. SALKIND, M. J. and V. Patarini. *Transactions*. American Institute of Mining, Metallurgical, and Petroleum Engineers, 239:1268 (1967).

25. SHIMMIN, K. D. and I. J. Toth. In *Failure Modes in Composites*. The Metallurgical Society of the American Institute of Mining, Metallurgical and Petroleum Engineers, p. 357 (1973).

26. DVORAK, G. J. and J. Q. Tarn. In *Fatigue of Composite Materials*, ASTM STP 569, American Society for Testing and Materials, p. 145 (1975).

27. SWANSON, G. D. and J. R. Hancock. In *Composite Materials: Testing and Design (Second Conference)*, ASTM STP 497, American Society for Testing and Materials, p. 469 (1972).

28. HANCOCK, J. R. In Technical Report AFFDL-TR-70-144. Air Force Flight Dynamics Laboratory, p. 285 (1970).

29. SCHEIRER, S. T. and I. J. Toth. Technical Report AFML-TR-73-178, Air Force Materials Laboratory (1973).

30. DHARAN, C. K. H. In *Fatigue of Composite Materials*, ASTM STP 569, American Society for Testing and Materials, p. 171 (1975).

31. DHARAN, C. K. H. *Journal of Materials Science*, 10:1665 (1975).

32. STURGEON, J. B. In *Proceedings of the 28th Annual Technical Conference*, The Society of the Plastics Industry, Inc., p. 12 (1973).

33. STURGEON, J. B. "Fatigue Testing of Carbon Fiber Reinforced Plastics," Technical Report 75135, Royal Aircraft Establishment, England (1975).

34. SHOCKEY, P. D., J. D. Anderson and K. E. Hofet. Technical Report AFML-TR-69-101, Air Force Materials Laboratory, Vol. V (1970).

35. HAHN, H. T. and W. K. Chin. To appear in *Composites Technology Review*.

36. CHIAO, C. C. and T. T. Chiao. "Aramid Fibers and Composites," Preprint UCRL-80400, Lawrence Livermore Laboratory (1977).

37. BUNSELL, A. R. *Journal of Materials Science*, 10:1300 (1975).

38. MORGAN, R. J., E. T. Mones, W. J. Steele and S. B. Deutscher. "The Structure and Property Relationships of Poly (p-Phenylene Terephthalamide) Fibers," Preprint UCRL-84208, Lawrence Livermore Laboratory (1980).

39. MORGAN, R. J., E. T. Mones and W. J. Steele. *Composites Technology Review*, 2(3) (1980).

40. HALL, D. *An Introduction to Composite Materials*. Cambridge University Press, NY (1985).

41. MAAS, D. R. *Mechanical Properties of Kevlar/SP 328*, CCM-83-19, Center for Composite Materials, University of Delaware (1983).

42. AWERBUCH, J. and H. T. Hahn. In *Fatigue of Filamentary Composite Materials*, ASTM STP 636, K. L. Reifsnider and K. N. Lauraitis, eds., American Society for Testing and Materials, p. 248 (1977).

43. HAHN, H. T. and R. Y. Kim. *Journal of Composite Materials*, 10:156 (1976).

44. HAENER, J., N. Ashbaugh, C. Y. Chia and M. Y. Feng. *Investigation of Micromechanical Behavior of Fiber Reinforced Plastics*, USAAVLABS-TR-66, AD-667901 (1968).

45. PREWO, K. M. and K. G. Kreider. In *Failure Modes in Composites*, The Metallurgical Society of the American Institute of Mining, Metallurgical and Petroleum Engineers, p. 395 (1973).

46. ADAMS, D. F. "The Enigma of the Eighties: Environment, Economics, Energy," *SAMPE Journal*, 24:1458 (1979).

47. MULLIN, J. V., V. F. Mazzio and R. L. Mehan. NASA CR-121000 2420-NO3, National Aeronautics and Space Administration (1972).

48. GOAN, J. C., T. W. Martin and R. Prescott. In *Proceedings of the 28th Technical Conference*, SPI, Inc., 21-B (1973).

49. TOTH, I. J., W. D. Brentnall and G. D. Menke. *Journal of Metals*, p. 71 (1972).

50. HAHN, H. T. and R. Y. Kim. *Journal of Composite Materials*, 10:156 (1976).

51. CHIAO, T. T. *Some Interesting Mechanical Behavior of Fiber Composite Materials*, Preprint UCRL-80908, Lawrence Livermore Laboratory (1978).

52. MENKE, G. D. and I. J. Toth. Technical Report AFML-TR-70-74, Air Force Materials Laboratory (1970).

53. LIFSCHITZ, J. M. In *Composite Materials, Vol. 5, Fracture and Fatigue*, L. J. Broutman, ed., Academic Press, NY, p. 124 (1974).

54. MORRIS, D. H. and H. T. Hahn. *Journal of Composite Materials*, 11:124 (1977).

55. CHIAO, T. T., C. C. Chiao and R. J. Sherry. In *Fracture Mechanics and Technology, Vol. 1*. G. C. Sih and C. L. Chow, eds., Sijthoff & Noordhoff, The Netherlands, p. 257 (1977).

56. WAGNER, H. D., P. Schwartz and S. L. Phoenix. "Lifetime Statistics for Single Kevlar 49 Filaments in Creep Rupture," *Journal of Materials Science*, 21:1868 (1986).

57. HAHN, H. T. and T. T. Chiao. In *Advances in Composite Materials, Vol. 1*. A. R. Bunsell et al., eds., Pergamon Press, p. 584 (1980).

58. Kevlar 49 Data Manual, E. I. du Pont de Nemours Chemical Co., Wilmington, DE (1974).

59. PENN, L. and F. Larson. *Journal of Applied Polymer Science*, 23:59 (1979).

60. GOUDIN, C. in *Advances in Composite Materials, Vol. 1*. A. R. Bunsell et al., eds., Pergamon Press, p. 497 (1980).

61. HAHN, H. T. and T. L. Gates. *Composites Technology Review*, 1(4):12 (1979).

62. JOHNSON, W. S. *Fatigue Testing and Damage Development in Continuous Fiber Reinforced Metal Matrix Composites*, NASA TM 100628 (June 1988).

63. HOFER, K. E., JR., D. Larsen and V. E. Humphreys. Technical Report AFML-TR-72-205, Air Force Materials Laboratory (1972).

64. HASHIN, Z. and A. Rotem. *Journal of Composite Materials*, 7:448 (1973).

65. SIMS, D. F. and V. H. Brogden. In *Fatigue of Filamentary Composite Materials*, ASTM STP 636, K. L. Reifsnider and K. N. Lauraitis, eds., p. 185 (1977).

66. (a) CHRISTIAN, J. L. In *Fatigue of Composite Materials*, ASTM STP 569, American Society for Testing and Materials, p. 280 (1975). (b) YANG, J. N. and D. L. Jones. *Journal of Composite Materials*, 12:371 (1978).

67. (a) HOFER, K. E., JR., D. Larsen and V. E. Humphreys. Technical Report AFML-TR-74-266, Air Force Materials Laboratory (1975). (b) NEVADUNSKY, J. J., J. J. Lucas and M. J. Salkind. *Journal of Composite Materials*, 9:394 (1975).

68. PIPES, R. B. In *Composite Materials: Testing and Design (Third Conference)*, ASTM STP 546, American Society for Testing and Materials, p. 419 (1972).

69. CHAMIS, C. C. and J. M. Sinclair. *Experimental Mechanics*, 17:248 (1977).

70. WHITNEY, J. M., I. M. Daniel and R. B. Pipes. *Experimental Mechanics of Fiber Reinforced Composite Materials*. rev. ed. (Society for Experimental Materials), Prentice-Hall, Inc., Englewood Cliffs, NJ (1984).

71. AWERBUCH, J. and H. T. Hahn. In *Fatigue of Fibrous Composite Materials*, ASTM STP 723, American Society for Testing and Materials, p. 243 (1981).

72. WU, E. M. In *Mechanics of Composite Materials, Vol. 2*. G. P. Sendeckyj, ed., Academic Press, NY (1974).

73. SANDHU, R. S. Technical Report AFFDL-TR-72-71, Air Force Flight Dynamics Laboratory (1972).

74. TSAI, S. W. and H. T. Hahn. Technical Report AFML-TR-77-33, Air Force Materials Laboratory (1977).

75. HAHN, H. T. *Journal of Composite Materials*, 9:316 (1975).

76. PIPES, R. B. and N. J. Pagano. *Journal of Composite Materials*, 4:538 (1970).

77. McGARRY, F. J. In *Fundamental Aspects of Fiber Reinforced Plastic Composites*. R. T. Schwartz and H. S. Schwartz, eds., Interscience, NY, p. 63 (1968).

78. BROUTMAN, L. J. and S. Sahu. In *Proceedings of the 24th Technical Conference*, SPI, Inc., 11-D (1969).

79. HAHN, H. T. and S. W. Tsai. *Journal of Composite Materials*, 8:288 (1974).

80. TANIMOTO, T. and S. Amijima. *Journal of Composite Materials*, 9:380 (1975).

81. GRIMES, G. C. and P. H. Francis. Technical Report AFML-TR-75-33, Air Force Materials Laboratory (1975).

82. REIFSNIDER, K. L., E. G. Henneke and W. W. Stinchcomb. Technical R Reports AFML-TR-76-81, Parts I–IV, Air Force Materials Laboratory (1976–1979).

83. DVORAK, G. J. and N. Laws. "Analysis of First Ply Failure in Composite Laminates," *Engineering Fracture Mechanics*, 25:763 (1986).

84. TSAI, S. W. and H. T. Hahn. In *Inelastic Behavior of Composite Materials*, AMD-Vol. 13, C. T. Herakovich, ed., ASME, p. 73 (1975).

85. KIM, R. Y. in *Advances In Composite Materials, Vol. 2*. A. R. Bunsell et al., eds., Pergamon Press, p. 1015 (1980).

86. HAHN, H. T. and D. G. Hwang. Technical Report AFWAL-TR-8-4172, Air Force Wright Aeronautical Laboratories (1980).

87. *Advanced Composites Design Guide*, Air Force Flight Dynamics Laboratory (1977).

88. RYDER, J. T. and E. K. Walker. Technical Report AFML-TR-76-241, Air Force Materials Laboratory (1976).

89. WHITNEY, J. M. Technical Report AFML-TR-79-4111, Air Force Materials Laboratory (1979).

90. KIM, R. Y. and W. J. Park. *Journal of Composite Materials*, 14:69 (1980).

91. SCHUTZ, D. and I. J. Gerharz. *Composites*, 8:245 (1977).

92. RAMANI, S. V. and D. P. Williams. In *Fatigue of Filamentary Composite Materials*, ASTM STP 636, K. L. Reifsnider and K. N. Lauraitis, eds., American Society for Testing and Materials, p. 27 (1977).

93. BROUTMAN, L. J. and S. Sahu. In *Composite Materials: Testing and Design (Second Conference)*, ASTM STP 497, American Society for Testing and Materials, p. 170 (1972).

94. LANG, R. W. and J. A. Manson. "Crack Tip Heating in Short Fiber Composites under Fatigue Loading Conditions," *Journal of Materials Science*, 22:3576 (1987).

95. DALLY, J. W. and L. J. Broutman. *Journal of Composite Materials*, 1:424 (1967).

96. SUN, C. T. and W. S. Chan. In *Composite Materials: Testing and Design (Fifth Conference)*, ASTM STP 674, S. W. Tsai, ed., American Society for Testing and Materials, p. 418 (1979).

97. CESSNA, L. C., J. A. Levens and J. B. Thomson. In *Proceedings of the 24th Annual Technical Conference*, SPI, Inc., 1-C (1969).

98. BEARDMORE, P. In *Fatigue Mechanisms*, ASTM STP 676, J. T. Fong, ed., American Society for Testing and Materials, p. 453 (1979).

99. REIFSNIDER, K. L., W. W. Stinchcomb, R. Williams and L. A. Marcus. Technical Report AFOSR-TR-73-1961, VPI&SU (1973).

100. WHITCOMB, J. D. In *Composite Materials: Testing and Design (Fifth Conference)*, ASTM STP 674, S. W. Tsai, ed., American Society for Testing and Materials, p. 502 (1979).

101. TSAI, G. C., J. F. Doyle and C. T. Sun. "Frequency Effects of the Fatigue Life and Damage of Graphite/Epoxy Composites," *J. Composite Materials*, 21:2 (1987).

102. SENDECKYJ, G. P. and H. D. Stalnaker. In *Composite Materials: Testing and Design (4th Conference)*, ASTM STP 617, American Society for Testing and Materials, p. 39 (1977).

103. CHANG, F. H., D. E. Gordon, B. T. Rodini and R. H. McDaniel. *Journal of Composite Materials*, 10:182 (1976).

104. CHANG, F. H., D. E. Gordon and A. H. Gardener. In *Fatigue of Filamentary Composite Materials*, ASTM STP 636, K. L. Reifsnider and K. N. Lauraitis, eds., American Society for Testing and Materials, p. 57 (1977).

105. UNDERWOOD, J. H. and D. P. Kendall. In *Proceedings of the 1975 International Conference on Composite Materials*, Vol. 2. The Metallurgical Society of the American Institute of Mining, Metallurgical and Petroleum Engineers, p. 1122 (1976).

106. DURCHLAUB, E. and P. Sacco. Technical Report AFFDL-TR-72-82, Air Force Flight Dynamics Laboratory (1972).

107. WADDOUPS, M. E., J. R. Eisenmann and B. E. Kaminski. *Journal of Composite Materials*, 5:446 (1971).

108. WALTER, R. W., R. W. Johnson, R. R. June and J. E. McCarty. In *Fatigue of Filamentary Composite Materials*, ASTM STP 636, K. L. Reifsnider and K. N. Lauraitis, eds., American Society for Testing and Materials, p. 228 (1977).

109. TSANGARAKIS, N. "The Notch-Fatigue Behavior of an Aluminum Composite Reinforced Unidirectionally with Silicon Carbide Fiber," *Journal of Composite Materials*, 21:1008 (1987).

110. DIAMANTAKOS, C. and R. J. Fritz. "Tensile Fatigue of Notched Carbon/Epoxy Specimens—Search for Optimum Model," *Journal of Reinforced Plastics and Composites*, 7:165 (1988).

111. GERHARZ, J. J. and D. Schutz. In *Review of Investigations of Aeronautical Fatigue in the Federal Republic of Germany*, ICAF Conference 1979-LBF Report S-142, Laboratorium für Betriebsfestigkeit, Darmstadt, Germany.

112. ROSENFELD, M. S. and S. L. Huang. Presented at the AIAA/ASME Structures, Dynamics and Materials Conference, San Diego (1977).

113. BEVAN, L. G. *Composites*, 8:227 (1977).

114. RATWANI, M. M. and H. P. Kan. "Effect of Stacking Sequence on Damage Propagation and Failure Modes in Composite Laminates," *Damage in Composite Materials*, ASTM STP 775, p. 211 (1982).

115. MOHLIN, T., A. F. Blom, L. A. Carlsson and A. I. Gustavsson. "Delamination Growth in a Notched Graphite/Epoxy Laminate under Compression Fatigue Loading," *Delamination and Debonding of Materials*, ASTM STP 876, p. 168 (1985).

116. MANDELL, J. F. and U. R. S. Meier. In *Fatigue of Composite Materials*, ASTM STP 569, American Society for Testing and Materials, p. 28 (1975).

117. KULKARNI, S. V., P. V. McLaughlin, Jr., R. B. Pipes and B. W. Rosen. In *Composite Materials: Testing and Design (Fourth Conference)*, ASTM STP 617, American Society for Testing and Materials, p. 70 (1977).

118. TSAI, S. W. and H. T. Hahn. *Introduction to Composite Materials*. Technomic Publishing Co., Inc., Lancaster, PA (1980).

119. WANG, A. S. D. In *Advances in Composite Materials, Vol. 1*. A. R. Bunsell et al., eds., Pergamon Press, p. 170 (1980).

120. PIPES, R. B. and N. J. Pagano. *Journal of Composite Materials*, 5:538 (1970).

121. WANG, A. S. D. and F. W. Crossman. *Journal of Composite Materials*, 11:92 (1977).

122. O'BRIEN, T. K., N. J. Johnston, D. H. Morris and R. A. Simonds. "A Simple Test for the Interlaminar Fracture Toughness of Composites," *SAMPE Journal*, 18:8 (1982).

123. O'BRIEN, T. K. "Characterization of Delamination Onset and Growth in a Composite Laminate," *Damage in Composite Materials*, ASTM STP 775, p. 140 (1982).

124. O'BRIEN, T. K. "Fatigue Delamination Behavior of PEEK Thermoplastic Composite Laminates," *Journal of Reinforced Plastics and Composites*, 7:341 (1988).

125. HAHN, H. T. *Journal of Composite Materials*, 10:266 (1976).

126. HAHN, H. T. and R. Y. Kim. In *Advanced Composite Materials—Environmental Effects*, ASTM STP 658, J. R. Vinson, ed., American Society for Composite Materials, p. 98 (1978).

127. SPRINGER, G. S. In *Composite Materials: Testing and Design (Fifth Conference)*, ASTM STP 674, S. W. Tsai, ed., American Society for Testing and Materials, p. 291 (1979).

128. BROWNING, C. E., G. E. Husman and J. M. Whitney. In *Composite Materials: Testing and Design (Fourth Conference)*, ASTM STP 617, American Society for Testing and Materials, p. 481 (1977).

129. PLUEDDEMANN, P., ed. *Interfaces in Polymer Matrix Composites*. Academic Press, NY (1974).

130. MORTON, J., S. Kellas and S. M. Bishop. "Damage Characteristics in Notched Carbon Fiber Composites Subjected to Fatigue Loading—Environmental Effects," *Journal of Composite Materials*, 22:657 (1988).

131. WATANABE, M. In *Composite Materials: Testing and Design (Fifth Conference)*, ASTM STP 674, S. W. Tsai., ed., American Society for Testing and Materials, p. 345 (1979).

132. HAHN, H. T. and R. Y. Kim. *Journal of Composite Materials*, 9:297 (1975).

133. DURCHLAUB, E. C. and R. B. Freeman. Technical Report AFML-TR-73-225, Vols. I and II, Air Force Materials Laboratory (1974).

134. FRANCIS, P. H., D. E. Walrath, D. F. Sims and D. N. Weed. *Journal of Composite Materials*, 11:488 (1977).

135. OWEN, M. J. and J. R. Griffiths. *Journal of Materials Science*, 13:1521 (1978).

136. HALPIN, J. C., K. L. Jerina and T. A. Johnson. In *The Test Methods for High Modulus Fibers and Composites*, ASTM STP 521, American Society for Testing and Materials, p. 5 (1973).

137. PHOENIX, S. L. In *Composite Materials: Testing and Design: (Fifth Conference)*, ASTM STP 674, S. W. Tsai, ed., American Society for Testing and Materials, p. 455 (1979).

138. YANG, J. N. and D. L. Jones. In *Advances in Composite Materials, Vol. 1.* A. R. Bunsell et al., eds., Pergamon Press, p. 472 (1980).

139. COX, H. L. *British J. Appl. Phy.*, 3:72 (1952).

140. HOLISTER, G. S. and C. Thomas. *Fibre Reinforced Materials*, Elsevier, Amsterdam (1966).

141. ROSEN, B. W. In *Fiber Composite Materials*. American Society for Metals, p. 37 (1965).

142. OWEN, M. J. in *Composite Materials, Vol. 5, Fracture and Fatigue*, L. J. Broutman, ed., Academic Press, NY, p. 313 (1974).

143. OWEN, M. J. and R. J. Howe. *J. Phys: D, Appl. Phys.*, 5:1637 (1972).

144. OWEN, M. J. and R. G. Rose. *Mod. Plast.*, 47:130 (1970).

145. MANDELL, J. F. and B. L. Lee. "Matrix Cracking in Short Fiber Reinforced Composites under Static and Fatigue Loading," presented at the ASTM Symp. on Composite Materials: Testing and Design (6th Conf.), Phoenix (May 12–13, 1981).

146. MANDELL, J. F., D. D. Huang and F. J. McGarry. In *Proc. 36th Ann. Tech. Conf.*, RP/C Inst., SPI, Inc., 10-A (1981).

147. MANDELL, J. F., D. D. Huang and F. J. McGarry. *Polymer Composites*, 2:137 (1981).

148. SMITH, T. R. and M. J. Owen. *Int. R. P. Conf., 6th Brit. Plas. Fed.*, London, Paper 27 (1968).

149. CHRISTENSEN, R. M. and J. A. Rinde. *Polymer Sci. and Eng.*, 19:49 (1979).

150. WANG, S. S., H. Snemasu and E. S. M. Chim. "Analysis of Fatigue Damage Evolution and Associated Anisotropic Elastic Property Degradation in Random Short-Fiber Composite," *Journal of Composite Materials*, 21:1084 (1987).

151. OWEN, M. J. *Symposium on Short Fiber Reinforced Composite Materials* (to be an ASTM STP Volume), ASTM, Minneapolis (April 1980).

152. MANDELL, J. F. In *Composite Reliability*, ASTM STP 580, ASTM, p. 515 (1975).

153. HOWE, R. J. and M. J. Owen. In *Proc. 8th Int. R. P. Conf.*, British Plastics Federation, Brighton, Paper 21 (1972).

154. DiBenedetto, A. T. "Fatigue Behavior of Glass Fiber Reinforced Polybutyleneterephthalate," Contract No. DAAG-46-78-C-0027, Army Materials and Mechanics Res. Center (April 1980).

155. JINEN, E. "Accumulated Strain in Low Cycle Fatigue of Short Carbon-Fiber Reinforced Nylon 6," *Journal of Materials Sci.*, 21:435 (1986).

156. CESSNA, L. C., J. A. Levens and J. B. Thomson. In *Proc. 24th Tech. Ann. Conf. RP/C Inst.*, SPI, Inc., 1-C (1969).

157. LANG, R. W. and J. A. Manson. "Crack Tip Heating in Short-Fiber Composites under Fatigue Loading Conditions," *Journal of Materials Sci.*, 22:3576 (1987).

158. HEIMBUCH, R. A. and B. A. Sanders. In *Composite Materials in the Automobile Industry*. S. V. Kulkarni, C. H. Zweben and R. B. Pipes, eds., ASME, p. 111 (1978).

159. ROWE, E. H. and F. J. McGarry. *Proc. 35th Annual Tech. Conf., RP/C Institute*, SPI, Inc., 18-E (1980).

160. LEE, B. L. and F. H. Howard. *Proc. 36th Annual Tech. Conf., RP/C Institute*, SPI, Inc. (1981).

161. DENTON, D. L. in *Proc. 34th Annual Tech. Conf., RP/C Institute*, SPI, Inc., 11-F (1979).

162. PETIT, P. H. and M. E. Waddoups. *J. Composite Materials*, 3:2 (1969).
163. CHOU, S. C., O. Orringer and J. H. Rainey. *J. Composite Materials*, 11:71 (1977).
164. NUISMER, R. J. In *Advances in Composite Materials, Vol. 1*. A. R. Bunsell et al., eds., Pergamon Press, p. 436 (1980).
165. ROTEM, A. and Z. Hashin. *AIAA Journal*, 14:868 (1976).
166. HASHIN, Z. "Analysis of Stiffness Reduction of Cracked Cross-Ply Laminates," *Engineering Fracture Mechanics*, 25:771 (1986).
167. KIM, R. Y. In *Test Methods and Design Allowables for Fibrous Composites*. C. C. Chamis, ed., ASTM STP 734, American Society for Testing and Materials, p. 91 (1981).
168. ROTEM, A. and H. G. Nelson. In *Fatigue of Fibrous Composite Materials*, ASTM STP 723, American Society for Testing and Materials, p. 152 (1981).
169. COLEMAN, B. D. *J. Appl. Phys.*, 29:968 (1958).
170. HAVILAND, R. P. *Engineering Reliability and Long Life Design*. D. Van Nostrand Co., Princeton (1964).
171. CHOU, P. C. and R. Croman. *J. Composite Materials*, 12:177 (1978).
172. YANG, J. N. and D. L. Jones. In *Fatigue of Fibrous Composite Materials*, ASTM STP 723, American Society for Testing and Materials, p. 213 (1981).
173. YANG, J. N. *J. Composite Materials*, 12:19 (1978).
174. SENDECKYJ, G. P. In *Test Methods and Design Allowables for Fibrous Composites*, ASTM STP 734, C. C. Chamis, ed., American Society for Testing and Materials, p. 245 (1981).
175. THOMAS, D. R. and L. J. Bain. *Technometrics*, 11:805 (1969).
176. PARK, W. J. "Pooled Estimations of the Parameters on Weibull Distributions," AFML-TR-79-4112, Air Force Materials Laboratory (1979).
177. WOLFF, R. V. and G. H. Lemon. "Reliability Prediction for Composite Joints-Bonded and Bolted," AFML-TR−74-197, Air Force Materials Laboratory (1976).
178. WHITNEY, J. M. in *Fatigue of Fibrous Composite Materials*, ASTM STP 723, ASTM, p. 133 (1981).
179. COHEN, A. C. *Technometrics*, 7:579 (1965).
180. LLOYD, D. K. and M. Lipow. *Reliability: Management, Methods, and Mathematics*. Prentice-Hall, Inc., Englewood Cliffs, NJ (1962).
181. *CEMCAL: Composites Experimental Mechanics Calculation*. Technomic Publishing Co., Inc., Lancaster, PA (1988).

SECTION 1.5

Mechanics of Two-Dimensional Woven Fabric Composites

1.5 MECHANICS OF TWO-DIMENSIONAL WOVEN FABRIC COMPOSITES

T.-W. CHOU

1.5.1 Introduction

Textile structural composites are gaining increasing technological importance. Textile forms used as reinforcements for composites can be designed to accommodate a variety of manipulative requirements, including dimensional stability, subtle conformability and deep-draw shapeability. Woven fabrics that are essentially two-dimensional constructions exhibit good stability in the mutually orthogonal warp and fill directions. Woven fabrics also provide more balance properties in the fabric plane than unidirectional laminae; the bidirectional reinforcement in a single layer of a fabric gives rise to excellent impact resistance. The ease of handling and low fabrication cost have made fabrics attractive for structural applications. Triaxially woven fabrics, made from three sets of yarns which interlace at 60-degree angles, offer improved isotropy and higher in-plane shear rigidity. There are also other two-dimensional materials in the forms of knit fabric and weft-inserted warp knit constructions (see subsection 1.2.3). These materials offer a much wider range of form and behavior than woven fabrics. It is possible to design composites with considerable flexibility in performance from complete directional stability to engineered directional elongation. Another emerging area in textile composites is based upon three-dimensional integrated structural geometry. These materials can assume complex shapes and provide a high level of transverse shear strength and impact resistance.

The objective of this section is to review models of the thermo-mechanical behavior of two-dimensional woven fabric composites. The fabrics are composed of two sets of mutually orthogonal yarns of the same material and, hence, hybrid fabrics are not included. Here, the term *yarns* represents individual filaments, untwisted fiber bundles, twisted fiber bundles or rovings. Discussion of the basic characteristics of fabric materials can be found in subsection 1.2.3.

The presentations of this section begin with a brief introduction of the geometric patterns of two-dimensional fabrics, followed by an outline of the methodology of analysis of fabric composites. Three techniques for modelling the stiffness and strength properties of fabric composites are then presented and their applicabilities are examined. The final subsection discusses the thermal behavior of fabric composites. The work reported in this section is described in more detail in references [1–10].

1.5.2 Geometric Characteristics

An orthogonal woven fabric consists of two sets of interlaced yarns. The length direction of the fabric is known as the warp, and the width direction is referred to as the fill or weft. The various types of fabrics can be identified by the pattern of repeat of the interlaced regions as shown in Figure 1.5-1. Two basic geometrical parameters can be defined to characterize a fabric; n_{fg} denotes that a warp yarn is interlaced with every n_{fg}-th fill yarn, and n_{wg} denotes that a fill yarn is interlaced with every n_{wg}-th warp yarn. Here, we confine ourselves to non-hybrid fabrics and the case of $n_{wg} = n_{fg} = n_g$. Fabrics with $n_g \geq 4$ and where the interlaced regions are not connected are known as satin weaves. As defined by their n_g values, the fabrics in Figure 1.5-1 are termed plain weave ($n_g = 2$), twill weave ($n_g = 3$), 4-harness satin ($n_g = 4$), and 8-harness satin ($n_g = 8$). The regions in Figure 1.5-1 indicated by the dotted lines define the "unit cells" or the basic repeating regions for the different weaving patterns. It is also noted that the top sides of the fabrics in Figure 1.5-1 are dominated by the fill yarns, whereas the bottom sides are dominated by the warp yarns.

131

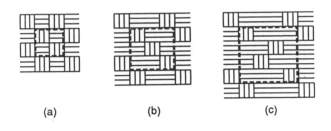

(a)   (b)   (c)

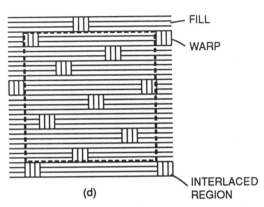

FILL

WARP

INTERLACED REGION

(d)

FIGURE 1.5-1. Examples of woven fabric patterns: (a) plain weave ($n_g = 2$); (b) twill weave ($n_g = 3$); (c) 4-harness satin ($n_g = 4$); and (d) 8-harness satin ($n_g = 8$).

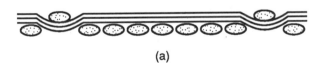

(a)

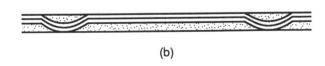

(b)

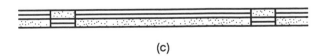

(c)

FIGURE 1.5-2. Idealization of the mosaic model: (a) cross-sectional view of a woven fabric before resin impregnation; (b) woven fabric composite; and (c) the idealization of the mosaic model.

1.5.3 Methodology of Analysis

The theoretical basis of the present analysis is the classical laminated plate theory [11,12]. Under the assumption of the Kirchoff-Love hypothesis, the constitutive equations are:

$$\begin{pmatrix} N_i \\ M_i \end{pmatrix} = \begin{bmatrix} A_{ij} & B_{ij} \\ B_{ij} & D_{ij} \end{bmatrix} \begin{pmatrix} \epsilon_j^0 \\ \varkappa_j \end{pmatrix} \quad (i,j = 1,2,6) \quad (1.5\text{-}1)$$

where N_i, M_i, ϵ_j^0, and $\varkappa_j$ indicate membrane stress resultants, moment resultants, strain and curvature of the laminate geometric midplane, respectively. The subscripts 1, 2, and 6 indicate, in the xyz coordinate system, the x direction, the y direction, and the xy plane, respectively. The components of the stiffness matrix A, B, and D are evaluated by integrating through the plate thickness in the z direction:

$$(A_{ij}, B_{ij}, D_{ij}) = \sum_{k=1}^{N} \int_{h_{k-1}}^{h_k} (l,z,z^2)Q_{ij}\,dz \quad (1.5\text{-}2)$$

Here, the lamina stiffness constants Q_{ij} corresponding to the lamina defined by h_k and h_{k-1} in the thickness direction are used in the calculations. More explicitly, Equation (1.5-2) can be written as:

$$A_{ij} = \sum_{k=1}^{N} (Q_{ij})_k (h_k - h_{k-1})$$

$$B_{ij} = \sum_{k=1}^{N} (Q_{ij})_k \frac{1}{2} (h_k^2 - h_{k-1}^2) \quad (1.5\text{-}3)$$

$$D_{ij} = \sum_{k=1}^{N} (Q_{ij})_k \frac{1}{3} (h_k^3 - h_{k-1}^3)$$

The inverted form of Equation (1.5-1) is given by:

$$\begin{pmatrix} \epsilon_i^0 \\ \varkappa_i \end{pmatrix} = \begin{bmatrix} a_{ij}^* & b_{ij}^* \\ b_{ij}^* & d_{ij}^* \end{bmatrix} \begin{pmatrix} N_j \\ M_j \end{pmatrix} \quad (1.5\text{-}4)$$

When the effect of temperature change is taken into

account, the constitutive relation of Equation (1.5-1) should be written as:

$$\begin{pmatrix} N_i \\ M_i \end{pmatrix} = \begin{bmatrix} A_{ij} & B_{ij} \\ B_{ij} & D_{ij} \end{bmatrix} \begin{pmatrix} \epsilon_j^0 \\ \varkappa_j \end{pmatrix} - \Delta T \begin{pmatrix} \tilde{A}_i \\ \tilde{B}_i \end{pmatrix} \quad (i,j = 1,2,6)$$

(1.5-5)

where

$$(\tilde{A}_i, \tilde{B}_i) = \sum_{m=1}^{N} \int_{h_{m-1}}^{h_m} (l,z) q_i \, dz \qquad (1.5-6)$$

$$q_i = Q_{ij} \alpha_j \qquad (1.5-7)$$

Here, ΔT indicates a small uniform temperature change, and α_j denotes the thermal expansion coefficient. q_i is evaluated for each lamina located between h_m and h_{m-1} in the thickness direction. After inversion, Equation (1.5-5) becomes:

$$\begin{pmatrix} \epsilon_i^0 \\ \varkappa_i \end{pmatrix} = \begin{bmatrix} a_{ij}^* & b_{ij}^* \\ b_{ij}^* & d_{ij}^* \end{bmatrix} \begin{pmatrix} N_j \\ M_j \end{pmatrix} + \Delta T \begin{pmatrix} \tilde{a}_j^* \\ \tilde{b}_j^* \end{pmatrix} \quad (1.5-8)$$

where

$$\begin{pmatrix} \tilde{a}_j^* \\ \tilde{b}_j^* \end{pmatrix} = \begin{bmatrix} a_{ij}^* & b_{ij}^* \\ b_{ij}^* & d_{ij}^* \end{bmatrix} \begin{pmatrix} \tilde{A}_i \\ \tilde{B}_i \end{pmatrix} \qquad (1.5-9)$$

$\tilde{a}_j^*$ and $\tilde{b}_j^*$ represent, respectively, the in-plane thermal expansion and thermal bending coefficients.

Based upon the iso-stress and iso-strain assumptions, the above constitutive equations can be used to obtain the bounds of the thermo-elastic properties [12]. The upper bounds of compliance constants are obtained from the iso-stress assumption; the lower bounds of stiffness constants are then obtained by taking the inversion of the compliance constant matrix. Similarly, the upper bounds of stiffness constants are derived from the iso-strain assumption. The lower bounds of compliance constants are then obtained by inverting the stiffness constant matrix.

1.5.4 The Mosaic Model

The basis of idealization of the "mosaic model" can be seen from Figure 1.5-2. Figure 1.5-2(a) is a cross-sectional view of an eight-harness satin. The consolidation of the fabric with a resin matrix material is depicted in Figure 1.5-2(b), which can be simplified to the mosaic model of Figure 1.5-2(c). The key simplification of the mosaic model is the omission of the fiber continuity and undulation (crimp) that exist in an actual fabric.

In general, a fabric composite idealized by the mosaic model can be regarded as an assemblage of pieces of asymmetric cross-ply laminates. Figure 1.5-3(a) shows the mosaic model of a unit cell for an eight-harness satin composite. The elastic stiffness constants of a cross-ply laminate [Figure 1.5-3(b)] can be derived based upon Equation (1.5-3). To this end, it is necessary to recapitulate the stiffness constants, Q_{ij}, of a unidirectional lamina which has orthotropic sym-

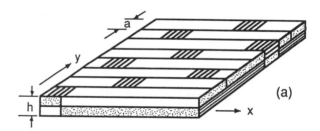

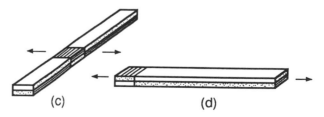

FIGURE 1.5-3. Mosaic model of: (a) repeating region in an eight-harness satin composite; (b) a basic cross-ply laminate; (c) parallel model; and (d) series model.

metry in the xy plane. Assuming that fibers are aligned along the x direction:

$$Q_{ij} = \begin{bmatrix} E_1/D_\nu & \nu_{12}E_2/D_\nu & 0 \\ \nu_{21}E_1/D_\nu & E_2/D_\nu & 0 \\ 0 & 0 & G_{12} \end{bmatrix} \quad (1.5\text{-}10)$$

where

$$D_\nu = 1 - \nu_{12}\nu_{21} \quad (1.5\text{-}11)$$

Here, E_1 and E_2 are the Young's moduli, G_{12} is the in-plane shear modulus, and ν_{12} denotes the Poisson's ratio relating the transverse strain in the y direction and the applied strain in the x direction. The Q_{ij} constants are symmetrical, i.e., $Q_{ij} = Q_{ji}$.

From Equations (1.5-3) and (1.5-10), the elastic stiffness constants of the cross-ply laminate shown in Figure 1.5-3(b) can be derived. The laminate is composed of two unidirectional laminae of thickness $h/2$. The total laminate thickness is h and the xy coordinate plane is positioned at the geometrical midplane of the laminate. Thus, $k = 1$ and 2 in Equation (1.5-3) and defines, respectively, laminae with fibers in the y and x directions. The non-vanishing stiffness constants are:

$$A_{11} = A_{22} = (E_1 + E_2)h/2D_\nu$$

$$A_{12} = \nu_{12}E_2h/D_\nu$$

$$A_{66} = G_{12}h$$

$$B_{11} = -B_{22} = (E_1 - E_2)h^2/8D_\nu \quad (1.5\text{-}12)$$

$$D_{11} = D_{22} = (E_1 + E_2)h^3/24D_\nu$$

$$D_{12} = \nu_{12}E_2h^3/12D_\nu$$

$$D_{66} = G_{12}h^3/12$$

The bending-stretching coupling constants B_{11} and B_{22} do not vanish because $E_1 \neq E_2$. Also, it is understood that A_{ij}, B_{ij}, and D_{ij} are symmetrical constants.

Using Equation (1.5-12), Equation (1.5-1) can be written in the following explicit form:

$$\begin{bmatrix} N_1 \\ N_2 \\ N_6 \end{bmatrix} = \begin{bmatrix} A_{11} & A_{12} & 0 \\ A_{12} & A_{11} & 0 \\ 0 & 0 & A_{66} \end{bmatrix} \begin{bmatrix} \epsilon_1^0 \\ \epsilon_2^0 \\ \epsilon_6^0 \end{bmatrix} + \begin{bmatrix} B_{11} & 0 & 0 \\ 0 & -B_{11} & 0 \\ 0 & 0 & 0 \end{bmatrix} \begin{bmatrix} \varkappa_1 \\ \varkappa_2 \\ \varkappa_6 \end{bmatrix}$$

$$(1.5\text{-}13)$$

$$\begin{bmatrix} M_1 \\ M_2 \\ M_6 \end{bmatrix} = \begin{bmatrix} B_{11} & 0 & 0 \\ 0 & -B_{11} & 0 \\ 0 & 0 & 0 \end{bmatrix} \begin{bmatrix} \epsilon_1^0 \\ \epsilon_2^0 \\ \epsilon_6^0 \end{bmatrix} + \begin{bmatrix} D_{11} & D_{12} & 0 \\ D_{12} & D_{11} & 0 \\ 0 & 0 & D_{66} \end{bmatrix} \begin{bmatrix} \varkappa_1 \\ \varkappa_2 \\ \varkappa_6 \end{bmatrix}$$

Inverting Equation (1.5-13), we obtain:

$$\begin{bmatrix} \epsilon_1^0 \\ \epsilon_2^0 \\ \epsilon_6^0 \end{bmatrix} = \begin{bmatrix} a_{11}^* & a_{12}^* & 0 \\ a_{12}^* & a_{11}^* & 0 \\ 0 & 0 & a_{66}^* \end{bmatrix} \begin{bmatrix} N_1 \\ N_2 \\ N_6 \end{bmatrix} + \begin{bmatrix} b_{11}^* & 0 & 0 \\ 0 & -b_{11}^* & 0 \\ 0 & 0 & 0 \end{bmatrix} \begin{bmatrix} M_1 \\ M_2 \\ M_6 \end{bmatrix}$$

$$(1.5\text{-}14)$$

$$\begin{bmatrix} \varkappa_1 \\ \varkappa_2 \\ \varkappa_6 \end{bmatrix} = \begin{bmatrix} b_{11}^* & 0 & 0 \\ 0 & -b_{11}^* & 0 \\ 0 & 0 & 0 \end{bmatrix} \begin{bmatrix} N_1 \\ N_2 \\ N_6 \end{bmatrix} + \begin{bmatrix} d_{11}^* & d_{12}^* & 0 \\ d_{12}^* & d_{11}^* & 0 \\ 0 & 0 & d_{66}^* \end{bmatrix} \begin{bmatrix} M_1 \\ M_2 \\ M_6 \end{bmatrix}$$

In the bound approach, the two-dimensional extent of the fabric composite plate is simplified by considering two one-dimensional models where the pieces of cross-ply laminates are either in parallel or in series as shown in Figures 1.5-3(c) and 1.5-3(d). In the parallel model, a uniform state of strain, ϵ^0, and curvature, $\varkappa$, in the laminate midplane is assumed as a first approximation. For the one-dimensional repeating region of length $n_g a$, where a denotes the yarn width, an average membrane stress, $\bar{N}_1$, is defined as:

$$\bar{N}_1 = \frac{1}{n_g a} \int_0^{n_g a} N_1 \, dy$$

$$= \frac{1}{n_g a} \left[\int_0^a (A_{11}\epsilon_1^0 + A_{12}\epsilon_2^0 + B_{11}\varkappa_1) \, dy \right.$$

$$\left. + \int_a^{n_g a} (A_{11}\epsilon_1^0 + A_{12}\epsilon_2^0 + B_{11}\varkappa_1) \, dy \right]$$

$$= (A_{11}\epsilon_1^0 + A_{12}\epsilon_2^0) + \frac{1}{n_g a} [aB_{11}^T + (n_g a - a)B_{11}^L]\varkappa_1$$

$$= A_{11}\epsilon_1^0 + A_{12}\epsilon_2^0 + \left(1 - \frac{2}{n_g}\right)B_{11}^L \varkappa_1 \quad (1.5\text{-}15)$$

The factor $(1 - 2/n_g)$ appears because the terms B_{11} for the interlaced region (B_{11}^T) and non-interlaced region (B_{11}^L) have opposite signs, namely, $B_{11}^T = -B_{11}^L$. It is noted that B_{11}^L is derived for a cross-ply with the same configuration as in Figure 1.5-3(b), where the upper surface $(z > 0)$ shows fibers in the x direction. B_{11}^T is for a cross-ply obtained by exchanging the positions of the two laminae in Figure 1.5-3(b). Other average stress resultants can be written similar to Equation (1.5-15) for the uniform midplane strain, ϵ^0, and curvature, $\varkappa$. The moment resultant, $\bar{M}_1$, for example, is:

$$\bar{M}_1 = \frac{1}{n_g a} \int_0^{n_g a} M_1 \, dy$$

$$= D_{11}\varkappa_1 + D_{12}\varkappa_2 + \left(1 - \frac{2}{n_g}\right)B_{11}^L \epsilon_1^0 \quad (1.5\text{-}16)$$

Let $\bar{A}_{ij}$, $\bar{B}_{ij}$, and $\bar{D}_{ij}$ be the stiffness constant matrices relating the average stress resultant $\bar{N}$ and moment resultant $\bar{M}$ with ϵ^0 and $\varkappa$ and obtain:

$$\bar{A}_{ij} = A_{ij}$$

$$\bar{B}_{ij} = \left(1 - \frac{2}{n_g}\right)B_{ij}^L \quad (1.5\text{-}17)$$

$$\bar{D}_{ij} = D_{ij}$$

These components provide upper bounds to the stiffness constants of the fabric composite based upon the one-dimensional model. If these stiffness constants are inverted, lower bounds of the elastic compliance constants can be obtained. All the elastic stiffness constants A, B, and D are computed based upon the basic laminate where the top layer is the fill threads [Figure 1.5-3(b)].

In the series model, the disturbance of stress and strain near the interface of the interlaced region is neglected. Let the model be subjected to a uniform in-plane force, N_t, in the longitudinal direction. The assumption of constant stress leads to the definition of an average curvature. For instance, the average curvature, $\bar{\varkappa}_1$, along the x direction is:

$$\bar{\varkappa}_1 = \frac{1}{n_g a} \int_0^{n_g a} \varkappa_1 \, dx$$

$$= \frac{1}{n_g a} \left[\int_0^a b_{11}^* N_1 \, dx + \int_a^{n_g a} b_{11}^* N_1 \, dx \right]$$

$$= \frac{1}{n_g a} [ab_{11}^{*T} + a(n_g - 1)b_{11}^{*L}]N_1$$

$$= \left(1 - \frac{2}{n_g}\right)b_{11}^{*L}N_1 \quad (1.5\text{-}18)$$

It is also understood that the terms b_{11}^* for the interlaced region (b_{11}^{*T}) and noninterlaced region (b_{11}^{*L}) are equal and opposite in signs. Other average curvature and midplane strain expressions can be written similar to Equation (1.5-18) for uniformly applied N and M. Let $\bar{a}_{ij}^*$, $\bar{b}_{ij}^*$, and $\bar{d}_{ij}^*$ be the compliance constant matrices relating the average midplane strain, $\bar{\epsilon}^0$, and curvature, $\bar{\varkappa}$, with the stress resultant, N, and moment resultant, M. Thus:

$$\bar{a}_{ij}^* = a_{ij}^*$$

$$\bar{b}_{ij}^* = \left(1 - \frac{2}{n_g}\right)b_{ij}^{*L} \quad (1.5\text{-}19)$$

$$\bar{d}_{ij}^* = d_{ij}^*$$

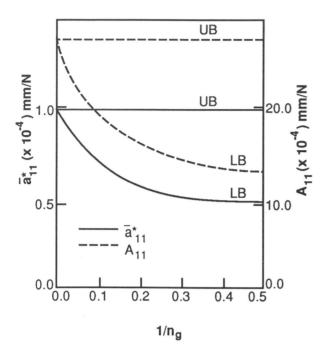

FIGURE 1.5-4. Variations of $\bar{a}^{*}_{11}$ and $\bar{A}_{11}$ with $1/n_g$.

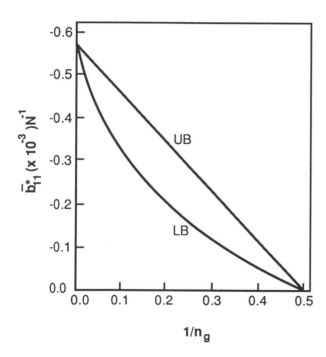

FIGURE 1.5-5. Variation of averaged coupling compliance with $1/n_g$.

Equation (1.5-19) gives the upper bounds of the composite compliance constants and, after inversion, the lower bounds of the stiffness constants.

In summary, both the upper and lower bounds of the elastic stiffness and compliance constants can be obtained from the mosaic model. Numerical results demonstrate the relationship between these bounds and $1/n_g$ are shown in Figures 1.5-4 and 1.5-5 for $\bar{A}_{11}(\bar{a}^{*}_{11})$ and $\bar{b}^{*}_{11}$, respectively. The following material properties of a graphite/epoxy composite given in Table 1.5-1 with fiber volume fraction in the impregnated yarn of 60% are adopted in the calculations. Unidirectional fiber composites are represented by the limiting case of $1/n_g \rightarrow 0$ ($n_g \rightarrow \infty$) and the upper and lower bounds of an elastic constant coincide with each. Plain weaves are represented by the case of $1/n_g = 0.5$. The coupling effects for plain weave composites vanish, as can be seen from Equations (1.5-17) and (1.5-19), and both the upper and lower bounds of $\bar{b}^{*}_{ij}$ ($\bar{B}_{ij}$) are identically zero. However, the bounds of $\bar{A}_{ij}$ ($\bar{a}^{*}_{ij}$) do not coincide for plain weave composites.

1.5.5 The Crimp (Fiber Undulation) Model

The crimp model is developed in order to consider the continuity and undulations of fibers in a fabric composite. Although the formulation of the problem developed in the following is valid for all n_g values, the crimp model is particularly suited for fabrics with low n_g values. The crimp model also provides the basis of analysis for the bridging model (subsection 1.5.6).

Figure 1.5-6 depicts the geometry of the model where the undulation shape is defined by the parameters $h_1(x)$, $h_2(x)$, and a_u. The parameters $a_0 = (a - a_u)/2$ and $a_2 = (a + a_u)/2$ are automatically determined by specifying a_u, which is geometrically arbitrary in the range from 0 to a. Because a pure matrix region appears in the model, an "overall" fiber volume fraction, V_f, can be different from V_f in the yarn region.

To simulate the actual configuration, the following form of crimp is assumed for the fill yarn:

$$h_1(x) = \begin{cases} 0 & (0 \leq x \leq a_0) \\ \left[1 + \sin\left\{\left(x - \dfrac{a}{2}\right)\dfrac{\pi}{a_u}\right\}\right]h_t/4 & (a_0 \leq x \leq a_2) \\ h_t/2 & (a_2 \leq x \leq n_g a/2) \end{cases}$$

$$(1.5\text{-}20)$$

The sectional shape of the warp yarn is expressed by:

$$h_2(x) = \begin{cases} h_t/2 & (0 \le x \le a_0) \\[2mm] \left[1 - \sin\left\{\left(x - \dfrac{a}{2}\right)\dfrac{\pi}{a_u}\right\}\right] \\[2mm] \qquad\qquad h_t/4 \ (a_0 \le x \le a/2) \\[2mm] -\left[1 + \sin\left\{\left(x - \dfrac{a}{2}\right)\dfrac{\pi}{a_u}\right\}\right]h_t/4 \\[2mm] \qquad\qquad (a/2 \le x \le a_2) \\[2mm] -h_t/2 & (a_2 \le x \le n_g a/2) \end{cases}$$

$$(1.5\text{-}21)$$

It is assumed that the laminated plate theory is applicable to each infinitesimal piece of the model along the x axis. Thus, A_{ij}, B_{ij}, and D_{ij} are expressed as functions of x $(0 \le x \le a/2)$ by:

$$A_{ij}(x) = \int_{-h/2}^{h_1(x)-h_t/2} Q_{ij}^M \, dz + \int_{h_1(x)-h_t/2}^{h_1(x)} Q_{ij}^F(\theta) \, dz$$

$$+ \int_{h_1(x)}^{h_2(x)} Q_{ij}^W \, dz + \int_{h_2(x)}^{h/2} Q_{ij}^M \, dz$$

$$= Q_{ij}^M[h_1(x) - h_2(x) + h - h_t/2]$$

$$+ Q_{ij}^F(\theta)h_t/2 + Q_{ij}^W[h_2(x) - h_1(x)]$$

$$B_{ij}(x) = \frac{1}{2} Q_{ij}^F(\theta)[h_1(x) - h_t/4]h_t \qquad (1.5\text{-}22)$$

$$+ \frac{1}{4} Q_{ij}^W[h_2(x) - h_1(x)]h_t$$

$$D_{ij}(x) = \frac{1}{3} Q_{ij}^M\{[h_1(x) - h_t/2]^3 - h_2(x)^3 + h^3/4\}$$

$$+ \frac{1}{3} Q_{ij}^F(\theta)[h_t^3/8 - 3h_t^2 h_1(x)/4 + 3h_t h_1^2(x)/2]$$

$$+ \frac{1}{3} Q_{ij}^W[h^2(x)^3 - h_1(x)^3]$$

where superscripts F, W, and M signify the fill yarn, warp yarn, and matrix, respectively. Similar expressions can be written for $a/2 \le x \le n_g a/2$.

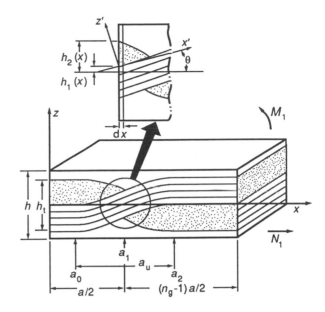

FIGURE 1.5-6. Fiber crimp model.

The local stiffness of the fill, $Q_{ij}^F(\theta)$, in the above equations is calculated as a function of the local off-axis angle, $\theta(x)$, which is defined as:

$$\theta(x) = \arctan\left(\frac{dh_1(x)}{dx}\right) \qquad (1.5\text{-}23)$$

Consider a fill yarn composed of parallel fibers. The fiber direction is denoted as the 1-direction; the 2- and 3-directions are perpendicular to the fiber and define the transversely isotropic plane. Then, from the

Table 1.5-1. *Material properties of unidirectional lamina.*

Material	Graphite/ Epoxy [1,5,13]		Glass/ Polyester [14]	Glass/ Polyimide [15,16]
Fiber volume fraction in impregnated yarns	60%	65%	60%	50%
E_1 (GPa)	113	132	47.5	41.2
E_2 (GPa)	8.82	9.31	15.9	15.7
G_{12} (GPa)	4.46	4.61	6.23	5.59
ν_{12}	0.3	0.28	0.27	0.3
ϵ_2^b	*	*	0.38%	0.5%
Thickness (mm)	0.4	0.4	0.4	0.244

*Strength calculations are not conducted.

Young's moduli (E_1, $E_2 = E_3$), shear moduli ($G_{12} = G_{13}$, G_{23}) and Poisson's ratio (ν_{12}) of the fill yarn, the elastic constants of the fill yarn with respect to the xyz axis in Figure 1.5-6 can be defined [17]. Here, the angle between the 1 and x axis is θ.

$$E_x^F(\theta) = 1/[\ell_\theta^4/E_1 + (1/G_{13} - 2\nu_{31}/E_1)\ell_\theta^2 m_\theta^2 + m_\theta^4/E_3]$$

$$E_y^F(\theta) = E_2 = E_3$$

$$G_{xy}^F(\theta) = 1/[\ell_\theta^2/G_{12} + m_\theta^2/G_{23}]$$ (1.5-24)

$$\nu_{yx}^F(\theta) = \nu_{31}\ell_\theta^2 + \nu_{23}m_\theta^2$$

where $\ell_\theta = \cos\theta$ and $m_\theta = \sin\theta$. It is also understood from the assumption of transverse isotropy of the fill yarn that $\nu_{12} = \nu_{13}$, $E_1/\nu_{12} = E_2/\nu_{21}$, $\nu_{23} = \nu_{32}$, and $G_{23} = E_2/2(1 + \nu_{23})$.

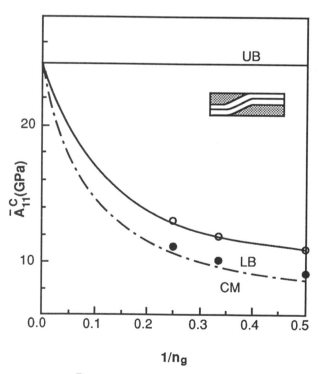

FIGURE 1.5-7. $\bar{A}_{11}^C$ against $1/n_g$ for graphite/epoxy, $V_f = 60\%$. Finite element results are indicated by $\bigcirc$ for the mosaic model and by $\bullet$ for the crimp model, —— mosaic model; —•—•— crimp model.

Thus, the local stiffness constants of the undulated portion of the fill yarn, referring to the xyz coordinate axes, are given as functions of fiber orientation θ:

$$Q_{ij}^F(\theta) = \begin{bmatrix} E_x^F(\theta)/D_\nu & E_y^F(\theta)\nu_{xy}^F(\theta)/D_\nu & 0 \\ E_x^F(\theta)\nu_{yx}^F(\theta)/D_\nu & E_y^F(\theta)/D_\nu & 0 \\ 0 & 0 & G_{xy}^F(\theta) \end{bmatrix}$$

$$i,j = 1,2,6$$ (1.5-25)

where

$$D_\nu = 1 - [\nu_{yx}^F(\theta)]^2 E_x^F(\theta)/E_y^F(\theta)$$ (1.5-26)

By substituting Equation (1.5-26) into Equation (1.5-22), the local plate stiffness constants can be evaluated. The local compliance constants, $a_{ij}^*(x)$, $b_{ij}^*(x)$, and $d_{ij}^*(x)$ are then obtained by inverting the stiffness constants $A_{ij}(x)$, $B_{ij}(x)$, and $D_{ij}(x)$.

We define the averaged in-plane compliance of the model under a uniformly applied in-plane stress resultant by:

$$\bar{a}_{ij}^{*C} = \frac{2}{n_g a} \int_0^{n_g a/2} a_{ij}^*(x) \, dx$$ (1.5-27)

where the superscript C signifies the fiber undulation model. Since $a_{ij}^*(x)$ is a constant within the straight portion of Figure 1.5-6, Equation (1.5-27) can be rewritten as:

$$\bar{a}_{ij}^{*C} = \left(1 - \frac{2a_u}{n_g a}\right) a_{ij}^* + \frac{2}{n_g a} \int_{a_0}^{a_2} a_{ij}^*(x) \, dx$$ (1.5-28)

where a_{ij}^* in the first term on the right-hand side of Equation (1.5-28) denotes the compliance of the straight portion of the yarns, namely, a cross-ply laminate and it is independent of x. The other compliance coefficients $\bar{b}_{ij}^{*c}$ and $\bar{d}_{ij}^{*c}$ are obtained in a similar manner:

$$\bar{b}_{ij}^{*c} = \left(1 - \frac{2}{n_g a}\right) b_{ij}^* + \frac{2}{n_g a} \int_{a_0}^{a_2} b_{ij}^*(x)\, dx$$

(1.5-29)

$$\bar{d}_{ij}^{*c} = \left(1 - \frac{2a_u}{n_g a}\right) d_{ij}^* + \frac{2}{n_g a} \int_{a_0}^{a_2} d_{ij}^*(x)\, dx$$

(1.5-30)

In the case of $n_g = 2$, $\bar{b}_{ij}^{*c}$ vanishes because that $b_{ij}^*(x)$ is an odd function with respect to $x = a/2$, the center of undulation, due to the assumed form of $h_1(x)$. Furthermore, Equations (1.5-28) to (1.5-30) coincide with the upper bounds of the compliance of Equation (1.5-19) as a_u tends to zero. The integrations in Equations (1.5-28) to (1.5-30) are conducted numerically because of the complexity of the integrands. The final results of the averaged elastic stiffness, $\bar{A}_{ij}^c$, $\bar{B}_{ij}^c$, and $\bar{D}_{ij}^c$, for the entire strip can be reached by inversion of $\bar{a}_{ij}^{*c}$, $\bar{b}_{ij}^{*c}$, and $\bar{d}_{ij}^{*c}$. If this procedure is applied in the warp direction, the balanced properties such as $\bar{A}_{11}^c = \bar{A}_{22}^c$ can be realized.

Numerical results demonstrating the relationship between the in-plane stiffness, $\bar{A}_{11}$, and $1/n_g$ are given in Figure 1.5-7. The unidirectional lamina properties of a graphite/epoxy system as shown in Table 1.5-1 are adopted. In Figure 1.5-7, UB and LB represent the results of the upper and lower bound predictions of the mosaic model; CM denotes the crimp model; circles indicate finite element results. Figure 1.5-7 demonstrates the reduction in $\bar{A}_{11}$ due to fiber undulation and the reduction is most severe in plain weave ($1/n_g = 0.5$) as compared to cross-ply laminates ($1/n_g = 0$).

The relationship between the coupling compliance $\bar{b}_{11}^*$ and $1/n_g$ is demonstrated in Figure 1.5-5. The results from the crimp model coincide exactly with those of the upper bound predictions because the second term on the right-hand side of Equation (1.5-29) vanishes due to the assumed asymmetrical shape of fiber undulation and, hence, the odd function representation of $b_{ij}^*(x)$ with respect to $x = a/2$.

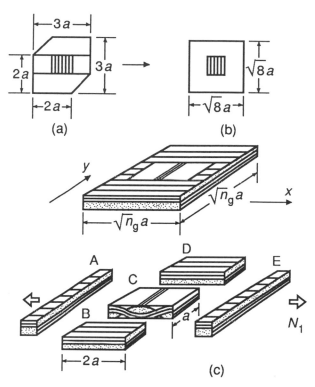

FIGURE 1.5-8. Concept of the bridging model. (a) Shape of the repeating unit of 8-harness satin; (b) modified shape for the repeating unit; and (c) idealization for the bridging model.

1.5.6 The Bridging Model

The success of the threadwise analysis has led to the concept of a bridging model for general satin composites. Such a model is desirable in view of the fact that the interlaced regions in a satin weave are separated from one another. The hexagonal shape of the repeating unit in a satin weave, as shown in Figure 1.5-8(a), is modified to a square shape [Figure 1.5-8(b)] for simplicity of calculation. A schematic view of the bridging model is shown in Figure 1.5-8(c) for a repeating unit which consists of the interlaced region and its surrounding areas. This model is valid only for satin weaves where $n_g \geq 4$. The four regions labelled by A, B, D, and E consist of straight fill yarns and hence can be regarded as pieces of cross-ply laminates of thickness h_t. Region C has an interlaced structure with an undulated fill yarn. Although the undulation and continuity in the warp yarns are ignored in this model, the effect is expected to be small because applied load is assumed to be in the fill direction.

The in-plane stiffness in region C where $n_g = 2$ has been calculated in subsection 1.5.5 and has been found to be much lower than that of a cross-ply laminate. Therefore, regions B and D carry higher loads than region C and act as bridges for load transfer between regions A and E. It is also assumed here that regions B, C, and D have the same averaged midplane strain and curvature. Then, the averaged stiffness constants for the regions B, C, and D are:

$$\bar{A}_{ij} = \frac{1}{\sqrt{n_g}} \left[(\sqrt{n_g} - 1)A_{ij} + \bar{A}_{ij}^C \right]$$

$$\bar{B}_{ij} = \frac{1}{\sqrt{n_g}} (\sqrt{n_g} - 1)B_{ij} \qquad (1.5\text{-}31)$$

$$\bar{D}_{ij} = \frac{1}{\sqrt{n_g}} \left[(\sqrt{n_g} - 1)D_{ij} + \bar{D}_{ij}^C \right]$$

$\bar{A}_{ij}^C$ and $\bar{D}_{ij}^C$ for the undulated portion C in Figure 1.5-8 are obtained from $\bar{a}_{ij}^{*c}$ and $\bar{d}_{ij}^{*c}$ of Equations (1.5-28) and (1.5-30), and $\bar{b}_{ij}^{*c} = 0$. A_{ij}, B_{ij}, and D_{ij} in Equation (1.5-31) for the cross-ply laminates of regions B and D in Figure 1.5-8 are given in Equation (1.5-12).

It is also postulated that the total in-plane force carried by regions B, C, and D is equal to that by region A or E. Then, the following averaged compliance constants are derived:

$$\bar{a}_{ij}^{*s} = \frac{1}{\sqrt{n_g}} \left[2\bar{a}_{ij}^* + (\sqrt{n_g} - 2)a_{ij}^* \right]$$

$$\bar{b}_{ij}^{*s} = \frac{1}{\sqrt{n_g}} \left[2\bar{b}_{ij}^* + (\sqrt{n_g} - 2)b_{ij}^* \right] \qquad (1.5\text{-}32)$$

$$\bar{d}_{ij}^{*s} = \frac{1}{\sqrt{n_g}} \left[2\bar{d}_{ij}^* + (\sqrt{n_g} - 2)d_{ij}^* \right]$$

where $\bar{a}_{ij}^*$, $\bar{b}_{ij}^*$, and $\bar{d}_{ij}^*$ are determined by inverting Equation (1.5-31) and the quantities with superscript S denote properties of the entire satin plane. Finally, $\bar{A}_{ij}^S$, $\bar{B}_{ij}^S$, and $\bar{D}_{ij}^S$ can be obtained by inverting Equation (1.5-32).

The reason why the fiber crimp model is effective for plain weave composites whereas the bridging model is valid for satin weave composites is explained below. There are no straight yarn regions surrounding an interlaced region in the plain weave. Moreover, the threadwise distribution of in-plane stiffness under the bending-free condition (see subsection 1.5.7) is identical in each yarn of a plain weave fabric in the loading direction. It can be expected, therefore, that no bridging effect occurs in the plain weave composite and that each yarn carries the same in-plane force. Hence, the threadwise analysis based on the fiber undulation model provides a reasonable prediction of the behavior of the plain weave composite.

Numerical results for the relationship between the in-plane elastic stiffness $\bar{A}_{11}^S$ and $1/n_g$ is indicated in Figure 1.5-9. The properties of the constitutive unidirectional laminae for a graphite/epoxy composite are given in Table 1.5-1. A prediction by the present theory shows an excellent agreement with experimental results of reference [18]. It should be noted that there is a slight drop of the overall fiber volume fraction due to the pure resin regions around the undulation. For instance, for a fiber volume fraction of 65%, the average overall fiber volume fraction in a repeating unit (Figure 1.5-8) for $n_g = 8$, $h_t = h$, and $a_u = a$ is around 62%.

1.5.7 Analysis of the Knee Behavior

Both the crimp model (subsection 1.5.5) and bridging model (subsection 1.5.6) described above are now extended to the study of the stress-strain behavior of woven fabric composites after initial fiber failure, known as a knee phenomenon [3]. The essential experimental fact for the knee phenomenon is that the breaking strain in the transverse layer, ϵ_2^b, is much smaller than that of the longitudinal layer in cross-ply laminates. Only the failure of the transverse yarns, which occurs in the warp direction in the present model, is considered. Thus, a failure criterion based upon maximum strain [12] is adopted.

In the following, we first apply the crimp model and confine our attention to the one-dimensional behavior of fabric composites under an applied force N_1. Then Equation (1.5-4) is reduced to:

$$\epsilon_1^0 = a_{11}^* N_1 + b_{11}^* M_1$$
$$\qquad (1.5\text{-}33)$$
$$\varkappa_1 = b_{11}^* N_1 + d_{11}^* M_1$$

where M_1 is the locally induced moment resultant due to the application of N_1. By assuming first that no bending deflection by the coupling effect is allowed along the x axis:

$$\varkappa_1 = b_{11}^* N_1 + d_{11}^* M_1 = 0 \qquad (1.5\text{-}34)$$

This assumption can be realized only if the fabric composite plate is symmetrical with respect to its midplane. In practical multilayer fabric composites arranged symmetrically to their midplanes, this assumption is expected to be approximately true. From Equations (1.5-33) and (1.5-34), we have:

$$\epsilon_1^0 = a_{11}^{**} N_1 \qquad (1.5\text{-}35)$$

where $a_{11}^{**} = a_{11}^* - b_{11}^{*2}/d_{11}^*$.

The quantity a_{11}^{**} may be referred to as a modified in-plane compliance and is a function of x. Since N_1 is uniform along the x-direction, $a_{11}^{**}(x)$ represents a strain distribution before the first internal failure. Figure 1.5-10 depicts two examples of the midplane strain distribution relative to that at the point $x = 0$ in Figure 1.5-6 and for $a_u = a$. It is easily seen that the fiber undulation causes local softening and that the maximum strain appears at the center of undulation ($x = a/2$). Also, the strain along the thickness direction at each section is uniform and equal to ϵ_1^0 owing to the classical plate theory and the bending free condition. Although the strain distribution calculated from finite element analysis [5] deviates slightly from the assumed uniform distribution, the present idealization provides a simple method for analyzing the knee phenomenon.

We consider that the highest strain in the region exceeds the specified strain ϵ_2^b first, and it immediately leads to the failure of the adjacent area. The damaged area in the warp yarn then propagates as the load increases. It is assumed that classical laminate theory is still valid during this failure process, and that the effective elastic moduli of such a failed area in the warp yarn are much lower than those of a sound area and can be expressed as:

$$Q_{ij}^{\prime W} = \begin{bmatrix} Q_{11}^W/100 & Q_{12}^W/100 & 0 \\ Q_{12}^W/100 & Q_{22}^W & 0 \\ 0 & 0 & Q_{66}^W/100 \end{bmatrix} \qquad (1.5\text{-}36)$$

Here, $Q_{ij}^{\prime W}$ denotes the reduced stiffness of the warp yarns after failure, and it is assumed that, with the ex-

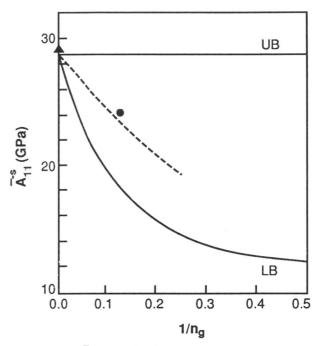

FIGURE 1.5-9. $\bar{A}_{11}^s$ vs. $1/n_g$ for a graphite/epoxy system, $V_f = 65\%$. —— upper and lower bounds; — — bridging model solution; ▲ and ●, experimental results for a cross-ply laminate and 8-harness satin, respectively [18].

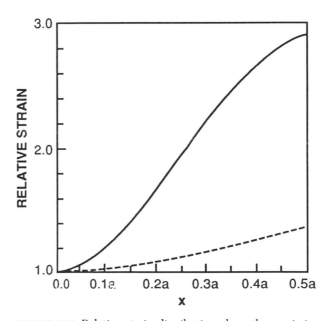

FIGURE 1.5-10. Relative strain distribution along the x-axis in the fiber crimp model under the bending-free condition, $a_u = a$. —— graphite/epoxy; — — glass/polyester.

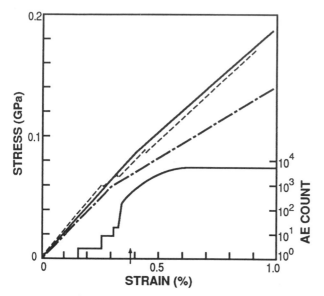

FIGURE 1.5-11. Stress-strain curves for plain weave composites of glass/polyester, $V_f = 36.8\%$ and experimental data of acoustic emission. —— presents results for the bending-free condition; — – — bending unconstrained condition; — — — finite element simulation [14]; ——— total count in acoustic emission measurement [14,19]; an arrow indicates the specified value of ϵ_2^b.

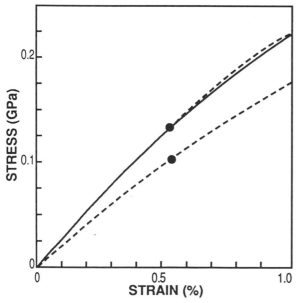

FIGURE 1.5-12. Theoretical and experimental stress-strain curves for a glass/polyimide composite, $V_f = 50\%$ in threads. ——— bridging model solution without bending for 8-harness satin (overall $V_f = 47.7\%$); — — — fiber undulation model solution without bending for plain weave (overall $V_f = 40.9\%$); — — an experimental curve from reference [15]; ● knee points.

ception of Q_{22}^W, the Q_{ij}'s are reduced by a factor of $1/100$ to reflect the weakening effect of transverse cracking [3]. The assumption of the applicability of the classical laminate theory implies that we neglect the complex stress and strain fields around the failed region. Such a successive failure process will continue until the lowest strain in the region reaches ϵ_2^b. At that time, all the warp regions are completely failed. Beyond this point, the stress-strain curve becomes a straight line again until the final failure of the fill yarns.

Next, we consider the case where the restraint on bending is removed. From the classical laminate theory:

$$\epsilon(z) = \epsilon_1^0 + z\varkappa_1 \qquad (1.5\text{-}37)$$

The strain state under an in-plane stress resultant, N_1, is given by:

$$\epsilon(z) = (a_{11}^* + zb_{11}^*)N_1 \qquad (1.5\text{-}38)$$

Thus, the strain field under the prescribed N_1 is determined from a_{11}^*, b_{11}^*, and z. Since the strain in a vertical section is distributed linearly according to Equation (1.5-37), it is necessary to determine the height, h_3, where the strain reaches the critical value, ϵ_2^b. If the strain at the outer edge of the warp yarns, $\epsilon_2(h_2)$ according to Equation (1.5-37), is larger than ϵ_2^b, we have, for $a_0 \leq x \leq a/2$:

$$h_3(x) = h_2 - (h_2 - h_1)\frac{\epsilon_2(h_2) - \epsilon_2^b}{\epsilon_2(h_2) - \epsilon_2(h_1)} \qquad (1.5\text{-}39)$$

By employing the h_3 value, the plate stiffness in Equation (1.5-22) needs to be modified after the initial failure. For instance, for $a_0 \leq x \leq a/2$:

$$A_{ij}(x) = Q_{ij}^M[h_1(x) - h_2(x) + h - h_t/2]$$
$$+ Q_{ij}^F(\theta)h_t/2 + Q_{ij}^W[h_3(x) - h_t(x)]$$
$$+ Q_{ij}'^W[h_2(x) - h_3(x)] \qquad (1.5\text{-}40)$$

Similar modifications to Equation (1.5-40) are made for B_{ij} and D_{ij} in Equation (1.5-22).

Figure 1.5-11 presents two numerical examples for a glass/polyester plain weave composite of $a_u = a$ and overall $V_f = 36.8\%$ with and without bending. The finite element analysis by other investigators [14] and their experimental result of acoustic emission [19] are also given. Basic material properties are shown in Table 1.5-1. The prediction for the bending-free condition compares very favorably with the finite element simulation. It is quite reasonable that the case with bending provides much lower stiffness because it is not subjected to lateral constraints.

In actual plain weave composites, local bending deformation caused by the coupling effect in each interlaced region is constrained by adjacent regions which have opposite signs of B_{ij}. Therefore, as far as plain weave composites are concerned, the one-dimensional analysis under the bending-free condition should give a reasonable prediction of the knee behavior under in-plane loading.

The bridging model and the concept of the successive warp yarn failure can be combined for analyzing the knee behavior in satin composites [3]. Similar to the approaches for plain weave composites, we consider two cases of bending. For the bending-free case:

$$\bar{A}_{11}^* = 1/\sqrt{n_g}\bar{A}_{11}^{*c} + (1 - 1/\sqrt{n_g})A_{11}^* \quad (1.5\text{-}41)$$

where $A_{11}^* = 1/a_{11}^{**}$ and a_{11}^{**} follows the definition in Equation (1.5-35). Due to the uniformity of N_1 along the x-direction, we obtain:

$$\bar{a}_{11}^{**s} = 2/\sqrt{n_g}\bar{a}_{11}^{**} + (1 - 2/\sqrt{n_g})a_{11}^{**}$$
$$\bar{A}_{11}^{*s} = 1/\bar{a}_{11}^{**s} \quad (1.5\text{-}42)$$

where $\bar{a}_{11}^{**} = 1/\bar{A}_{11}^*$. Similar expressions for the unconstrained bending case can also be obtained but are omitted here. The rest of the procedure for examining the knee phenomenon is quite similar to that of the plain weave case. The initial failure of the warp yarns occurs at the point of highest strain, for example, the center of the undulation in the bending free case. Also, since there are regions of uniform strains such as the bridging zones in this model, the entire area of those regions may fail simultaneously, according to the present assumptions.

Figure 1.5-12 compares numerical and experimental results for stress-strain curves of an 8-harness satin fabric plate of glass/polyimide composites. Basic material properties are also indicated in Table 1.5-1 where certain values are estimated from matrix data. The experimental curve is reproduced from reference [15]. Since test pieces were curved nearly symmetrically with respect to their midplanes, the bending free analysis is selected for comparison. It can be seen that the

agreement is quite good, particularly for strain values up to the point of the knee. A theoretical curve for the plain weave composite of the same material is also shown in Figure 1.5-12. We define a knee point by a deviation of 0.01% in strain from the linear strain. Then, we observe that knee stress in the 8-harness satin is higher than that of the plain weave although knee strains are nearly identical. It can be concluded that the elastic stiffness and knee stress in satin composites are higher than those in plain weave composites due to the presence of the bridging regions.

1.5.8 Summaries of the Stiffness and Strength Modellings of the Two-Dimensional Woven Fabric Composites

1. A fabric composite can be idealized as an assemblage of pieces of asymmetric cross-ply laminates. The upper and lower bounds of elastic stiffness and compliance of fabric composite plates in such a "mosaic model" are obtained based upon the constant strain and constant stress assumption.

2. The "crimp model," which is a one-dimensional approximation and takes into account fiber continuity and undulation, is particularly suited for predicting elastic properties of plain weave composites. The analytical results based upon the crimp model demonstrate that fiber undulation leads to a softening in the in-plane stiffness as compared to the mosaic model. However, fiber undulation has no effect on the coupling constants. Therefore, the solutions of the coupling compliance based upon the mosaic model can be considered to be reliable.

The results of both the crimp model and the mosaic model for the compliance constants $\bar{a}_{11}^*$ and $\bar{b}_{11}^*$ compare very favorably with the results of a finite element analysis [5].

3. In the case of $\bar{D}_{11}^*$, a transverse shear deformation theory is adapted for a modification of the mosaic model to examine the response of a fabric composite plate under both cylindrical bending and lateral force. Numerical results of $\bar{d}_{11}^*$ based upon the modified transverse shear deformation theory coincide well with the finite element results [5].

4. The effect of fiber undulation shapes on $\bar{a}_{11}^*$ in the crimp model has been examined—the result is shown in Figure 1.5-13. The geometrical parameters a, h, and a_u are chosen to be 1.0, 0.4, and 0.6 mm, respectively. The

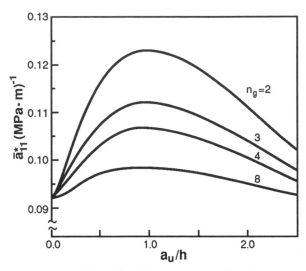

FIGURE 1.5-13. Relationships between averaged in-plane compliance and undulation length.

ter, they carry higher loads and play the role of load transferring bridges.

8. The initial elastic stiffness of satin composites can be predicted by the bridging model. The present analysis of an 8-harness satin composite of graphite/epoxy demonstrates good agreement with experimental data for the fabric and for a cross-ply laminate as the limiting case.

9. The concepts of successive failure of the warp yarns and the bridging idealization have been combined to study the knee behavior in satin composites. The theoretical results for an 8-harness satin reinforced glass/polyimide composite compares extremely well with the experimental curve. It has been concluded that the bridging regions surrounding the interlaced regions are responsible for the higher stiffness and knee stress in satin composites than those in plain weave composites.

calculations are performed for the range of a_u/h values from 0 to a/h, where the case $a_u \to 0$ corresponds to the configuration of a mosaic model. The results show that $\bar{a}_{11}^*$ is susceptible to the undulation shape, particularly at small n_g values. The highest $\bar{a}_{11}^*$ value, i.e., the lowest in-plane stiffness, is obtained at around $a_u/h = 1$. On the other hand, the $\bar{a}_{11}^*$ values at $a_u/h = 0$ and a/h are not far apart. Because in actual fabrics a_u may be fairly close to a and hence $a_u/h = a/h$, the mosaic model ($a_u/h = 0$) seems to be a convenient model to evaluate the in-plane stiffness of a fabric.

5. The crimp model has been applied to examine the knee failure phenomenon of plain weave composites. The predicted knee behavior of a glass/polyester composite under the bending-free assumption shows an excellent agreement with the stress-strain curve obtained by using a finite element analysis.

6. The bound theory based upon the mosaic model is useful for a rough estimation of fabric composite stiffness properties. The crimp model offers a better predictability than the mosaic model for the in-plane and bending moduli. However, the crimp model is inadequate for evaluating the elastic properties of satin weave composites with large n_g.

7. A bridging model has been developed to examine the stiffness and strength of general satin composites. The interlaced regions in a satin fabric are separated from one another by the non-interlaced regions. Since the regions with straight yarns surrounding an interlaced region have higher in-plane stiffness than the lat-

1.5.9 In-Plane Thermal Expansion and Thermal Bending Coefficients

The constitutive equations of a laminated plate taking into account the effects due to a small uniform temperature change are given in Equations (1.5-5) to (1.5-9). In the following, the analytical techniques developed for the mosaic model (subsection 1.5.4), crimp model (subsection 1.5.5), and bridging model (subsection 1.5.6) are applied to analyze the thermal problem.

First, for applying the mosaic model, we again consider a long strip of the fabric composite as shown in Figure 1.5-3(a). The laminate is free of externally applied loading. The averaged strains and curvatures of a one-dimensional strip of width a in Figure 1.5-3(a) along the fill or warp direction due to a uniform temperature change, ΔT, can be expressed in the following several forms:

$$\bar{\epsilon}_i^0 = \frac{1}{n_g a} \int_0^{n_g a} \Delta T \tilde{a}_i^*(\xi) \, d\xi = \Delta T \tilde{a}_i^* \quad (i = 1,2)$$

$$(1.5\text{-}43)$$

$$\bar{\varkappa}_i = \frac{1}{n_g a} \int_0^{n_g a} \Delta T \tilde{b}_i^*(\xi) \, d\xi = \Delta T \frac{n_{g-2}}{n_g} \tilde{b}_i^* \quad (i = 1,2)$$

$$(1.5\text{-}44)$$

Here, ξ stands for x or y and the bar denotes average of a quantity. Because of the nature of the cross-ply laminates $\tilde{a}_6^*$ and $\tilde{b}_6^*$ vanish. From Equations (1.5-43) and (1.5-44), the average thermal expansion and thermal bending coefficients for the mosaic model are given by:

$$\bar{\tilde{a}}_i^* = \tilde{a}_i^*$$

$$\bar{\tilde{b}}_i^* = \left(1 - \frac{2}{n_g}\right)\tilde{b}_i^* \qquad (1.5\text{-}45)$$

Next, we apply the crimp model for taking into account the effect due to fiber undulation. The forms of fiber crimp for the fill and warp yarns follow the assumed shapes of Equations (1.5-20) and (1.5-21), respectively. Let ξ again represent the x or y coordinate in Figure 1.5-6.

By assuming no in-plane force and moment and following the derivations of Equations (1.5-43) and (1.5-44), we obtain the fiber crimp model:

$$\bar{\tilde{a}}_i^{*C} = \left(1 - \frac{2a_u}{n_g a}\right)a_i^* + \frac{2}{n_g a}\int_{a_0}^{a_2} a_i^*(\xi)\,d\xi$$

$$(1.5\text{-}46)$$

$$\bar{\tilde{b}}_i^{*C} = \left(1 - \frac{2}{n_g}\right)\tilde{b}_i^* + \frac{2}{n_g a}\int_{a_0}^{a_2} \tilde{b}_i^*(\xi)\,d\xi$$

$$(1.5\text{-}47)$$

Here, the superscript C signifies the crimp model. It is understood that $\bar{\tilde{a}}_6^{*C}$ and $\bar{\tilde{b}}_6^{*C}$ vanish for cross-ply constructions. Since $\tilde{b}_i(\xi)$ is an odd function of ξ with respect to $\xi = a_1$ due to the assumed form of $h_1(\xi)$, the integration in Equation (1.5-47) vanishes and

$$\bar{\tilde{b}}_i^{*C} = \left(1 - \frac{2}{n_g}\right)\tilde{b}_i^* \qquad (1.5\text{-}48)$$

Equation (1.5-48) is identical to Equation (1.5-45) obtained from the mosaic model and this indicates that fiber crimp has no effect on the thermal bending coefficients. Identical conclusions were obtained for the bending-stretching coupling constant in subsection 1.5.5.

For the in-plane thermal expansion coefficient, it is necessary to evaluate the integration in Equation (1.5-46). This is done based upon the assumption that the classical laminated plate theory is applicable to each infinitesimal piece of width dx of the one-dimensional strip shown in Figure 1.5-6. The following steps are taken to obtain $\tilde{a}_i^*(\xi)$. First, $\tilde{A}_i(\xi)$ and $B_i(\xi)$ are evaluated from Equations (1.5-6) and (1.5-7) for $0 \le \xi \le a/2$ and the results are:

$$\tilde{A}_i(\xi) = q_i^M[h_1(\xi) - h_2(\xi) + h - h_t/2]$$
$$+ q_i^F(\theta)h_t/2 + q_i^W[h_2(\xi) - h_1(\xi)] \quad (1.5\text{-}49)$$

$$\tilde{B}_i(\xi) = \frac{1}{2} q_i^F(\theta)[h_t(\xi) - h_t/4]h_k$$

$$+ \frac{1}{4} q_i^W[h_2(\xi) - h_1(\xi)]h_t \qquad (1.5\text{-}50)$$

where the superscripts F, W, and M signify the fill yarn, warp yarn, and matrix regions, respectively. q_i is determined from Equation (1.5-7); $q_i^F(\theta)$, in particular, is determined from the local stiffness matrix $Q_{ij}^F(\theta)$, following the procedures outlined in subsection 1.5.5. Furthermore, the off-axis thermal expansion coefficients $\alpha_i^F(\theta)$ of Equation (1.5-7) are given by:

$$\alpha_1^F(\theta) = \alpha_L^F \cos^2\theta + \alpha_T^F \sin^2\theta$$

$$\alpha_2^F(\theta) = \alpha_T^F \qquad (1.5\text{-}51)$$

$$\alpha_6^F(0) = 0$$

where α_L and α_T denote, respectively, thermal expansion coefficients parallel and transverse to the fiber direction in a unidirectional fiber composite. Thus, $\tilde{A}_i(\xi)$ and $\tilde{B}_i(\xi)$ can be determined from Equations (1.5-49) and (1.5-50). Also, $a_{ij}^*(\xi)$, $b_{ij}^*(\xi)$, and $d_{ij}^*(\xi)$ in Equation (1.5-8) are obtained by inversion of $A_{ij}(\xi)$, $B_{ij}(\xi)$, and $D_{ij}(\xi)$. Finally, $\tilde{a}_i(\xi)$ can be derived from Equation (1.5-9).

Numerical integration of Equation (1.5-46) has been conducted and the results for $\bar{\tilde{a}}_1^{*C}$ and $\bar{\tilde{b}}_1^{*C}$ as functions of $1/n_g$ are given in Figure 1.5-14. It should be noted that the balanced thermal property such as $\bar{\tilde{a}}_1^* = \bar{\tilde{a}}_2^*$ for a fabric composite can be realized if the above procedure of calculation is conducted for one-directional strips along both the fill and warp directions.

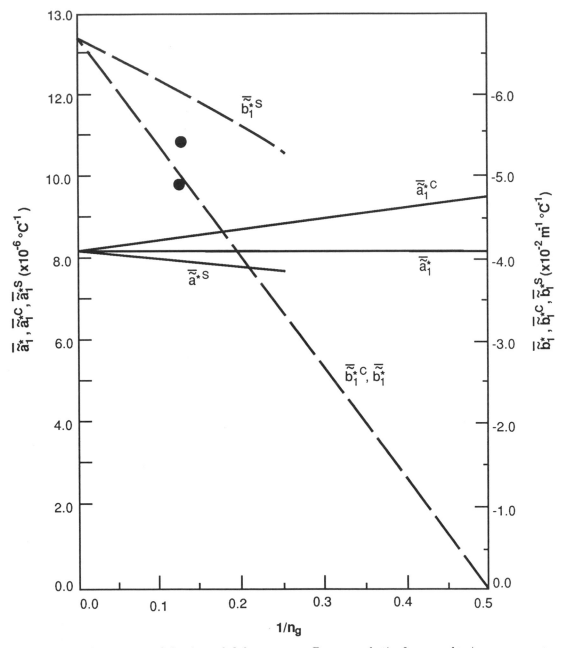

FIGURE 1.5-14. The variation of the thermal deformation coefficients with $1/n_g$ for a graphite/epoxy composite, $V_f = 60\%$ and $a_u/a = 1.0$. $\bar{\bar{a}}_i^*$: ———, $\bar{b}_i^*$: — — —, and ●: experimental results of b_1 at 300 K.

Lastly, the bridging model is applied to analyze the thermal properties. It has been noted in subsection 1.5.6 that the regions B and D of Figure 1.5-8 are stiffer than the crimped region C and, hence, they carry more load when an external force is applied in the x direction. Regions B, C, and D are termed bridging regions. For the thermal property analysis, assuming no external loading, the equilibrium of the bridging regions requires:

$$a \begin{Bmatrix} N_i^c \\ M_i^c \end{Bmatrix} + (\sqrt{n_g} - 1)\, a \begin{Bmatrix} N_i \\ M_i \end{Bmatrix} = 0 \quad (1.5\text{-}52)$$

where the superscript C again denotes the crimped region, and N_i and M_i without superscripts are for the cross-ply laminate. Furthermore, under the assumption of uniform strain and curvature in the bridging regions B, C, and D, we define:

$$\epsilon_i^{0C} = \bar{\epsilon}_i^0$$
$$\varkappa_i^C = \bar{\varkappa}_i \qquad (1.5\text{-}53)$$

where the bar denotes average of the bridging regions.

Substituting Equation (1.5-5) into Equation (1.5-52) and taking into account Equation (1.5-53), we have:

$$\left(\begin{bmatrix} A_{ij}^C & B_{ij}^C \\ B_{ij}^C & D_{ij}^C \end{bmatrix} + (\sqrt{n_g} - 1) \begin{bmatrix} A_{ij} & B_{ij} \\ B_{ij} & D_{ij} \end{bmatrix} \right) \begin{Bmatrix} \bar{\epsilon}_j^0 \\ \bar{\varkappa}_j \end{Bmatrix}$$

$$= \Delta T \left[\begin{Bmatrix} \tilde{A}_i^C \\ \tilde{B}_i^C \end{Bmatrix} + (\sqrt{n_g} - 1) \begin{Bmatrix} \tilde{A}_i \\ \tilde{B}_i \end{Bmatrix} \right] \qquad (1.5\text{-}54)$$

The quantities on the left-hand side of Equation (1.5-54) can be related to the average elastic stiffness in the bridging regions as:

$$\begin{bmatrix} A_{ij}^C & B_{ij}^C \\ B_{ij}^C & D_{ij}^C \end{bmatrix} + (\sqrt{n_g} - 1) \begin{bmatrix} A_{ij} & B_{ij} \\ B_{ij} & D_{ij} \end{bmatrix} = \sqrt{n_g} \begin{bmatrix} \bar{A}_{ij} & \bar{B}_{ij} \\ \bar{B}_{ij} & \bar{D}_{ij} \end{bmatrix}$$

$$(1.5\text{-}55)$$

Hence, Equation (1.5-54) can be rewritten as:

$$\begin{Bmatrix} \bar{\epsilon}_i^0 \\ \bar{\varkappa}_i \end{Bmatrix} = \Delta T \begin{bmatrix} \bar{a}_{ij}^* & \bar{b}_{ij}^* \\ \bar{b}_{ij}^* & \bar{d}_{ij}^* \end{bmatrix} \left(\frac{1}{\sqrt{n_g}} \begin{Bmatrix} \tilde{A}_j^C \\ \tilde{B}_j^C \end{Bmatrix} + \left(1 - \frac{1}{\sqrt{n_g}} \right) \begin{Bmatrix} \tilde{A}_j \\ \tilde{B}_j \end{Bmatrix} \right)$$

$$(1.5\text{-}56)$$

Here, $\bar{a}_{ij}^*$, $\bar{b}_{ij}^*$, and $\bar{d}_{ij}^*$ are obviously obtained by inverting $\bar{A}_{ij}$, $\bar{B}_{ij}$, and $\bar{D}_{ij}$. In comparison with Equation (1.5-8) the quantities in the parentheses on the right-hand side of Equation (1.5-56) can be regarded as the

average values for the bridging regions and, hence, they are denoted by $\bar{\bar{A}}_j$ and $\bar{\bar{B}}_j$. Thus, we obtain:

$$\begin{Bmatrix} \bar{\bar{a}}_i^* \\ \bar{\bar{b}}_i^* \end{Bmatrix} = \begin{bmatrix} \bar{a}_{ij}^* & \bar{b}_{ij}^* \\ \bar{b}_{ij}^* & \bar{d}_{ij}^* \end{bmatrix} \begin{bmatrix} \bar{\bar{A}}_j \\ \bar{\bar{B}}_j \end{bmatrix} \qquad (1.5\text{-}57)$$

Finally, the whole satin composite of Figure 1.5-8 can be regarded as a linkage of the regions A, B-C-D, and E in series. The average strains and curvature for the entire model are given by:

$$\begin{Bmatrix} \bar{\epsilon}_i^{0S} \\ \bar{\varkappa}_i^S \end{Bmatrix} = \frac{1}{\sqrt{n_g}} \left(2 \begin{Bmatrix} \bar{\epsilon}_i^0 \\ \bar{\varkappa}_i \end{Bmatrix} + (\sqrt{n_g} - 2) \begin{Bmatrix} \epsilon_i^0 \\ \varkappa_i \end{Bmatrix} \right)$$

$$= \Delta T \left(\frac{2}{\sqrt{n_g}} \begin{Bmatrix} \bar{\bar{a}}_i^* \\ \bar{\bar{b}}_i^* \end{Bmatrix} + \left(1 - \frac{2}{\sqrt{n_g}} \right) \begin{Bmatrix} \tilde{a}_i^* \\ \tilde{b}_i^* \end{Bmatrix} \right)$$

$$(1.5\text{-}58)$$

where the superscript S signifies the properties of the satin composite and ϵ^0 and $\varkappa$ denote midplane strain and curvature for the cross-plies in regions A and E of Figure 1.5-8. Equation (1.5-58) implies

$$\begin{Bmatrix} \bar{\bar{a}}_i^{*S} \\ \bar{\bar{b}}_i^{*S} \end{Bmatrix} = \frac{2}{\sqrt{n_g}} \begin{Bmatrix} \bar{\bar{a}}_i^* \\ \bar{\bar{b}}_i^* \end{Bmatrix} + \left(1 - \frac{2}{\sqrt{n_g}} \right) \begin{Bmatrix} \tilde{a}_i^* \\ \tilde{b}_i^* \end{Bmatrix} \qquad (1.5\text{-}59)$$

for the thermal expansion and thermal bending coefficients of the satin composites.

Figure 1.5-14 shows numerical results of the analysis. The elastic properties of constituent unidirectional laminae are the same as those in Table 1.5-1 for a graphite/epoxy composite with a fiber volume fraction of 60%. Also, $\alpha_1 = 0.0$ and $\alpha_2 = 3.0 \times 10^{-5}\,°\text{C}^{-1}$. The general characteristics of the variations of thermal deformation coefficients with $1/n_g$ are very similar to those of the compliance constants $\bar{a}_{11}^*$ and $\bar{b}_{11}^*$ as discussed earlier. For the thermal bending coefficients, there is considerable discrepancy between the results

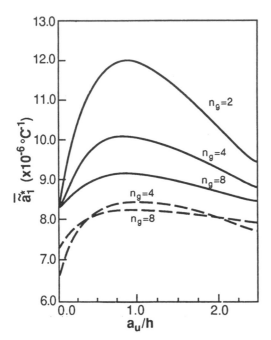

FIGURE 1.5-15. The effect of fiber undulation on $\bar{\tilde{a}}_1^*$, crimp model: ———; bridging model: — — —.

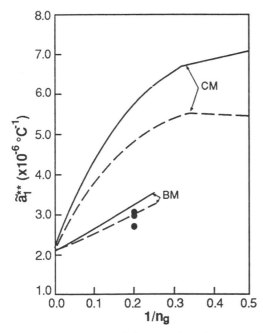

FIGURE 1.5-16. Comparison of theoretical predictions with the experimental results of reference [20] for 5-harness satin graphite/epoxy composites. $a/h = 3.75$: ———, $a/h = 7.5$: — — —, and $a_u/a = 1.0$. CU and BM indicate fiber crimp and bridging models, respectively. ●: experimental results at 300 K.

obtained from the one-dimensional models and the bridging model. Although the experimental results of reference [1] are closer to the one-dimensional predictions, the data are insufficient to provide a valid comparison with the theoretical models.

The geometrical shape of the fiber undulation which also affects $\bar{\tilde{a}}_1^*$ is demonstrated in Figure 1.5-15 where material properties of Table 1.5-1 also have been used. The results indicate that the in-plane thermal expansion coefficient of satin weave composites is less sensitive to a_u/h than that of plain weave composites. Furthermore, the fiber crimp model predicts a larger effect on $\bar{\tilde{a}}_1^*$ due to a_u/h than the bridging model. In general, the bridging model predictions are also less sensitive to the n_g values than the crimp model. In both models, the maximum in $\bar{\tilde{a}}_1^*$ occurs at $a_u/h = 1$.

Experimental data on thermal expansion coefficients of fabric composites are extremely limited. References [20–23] present measurements of thermal expansion of graphite and glass fiber composites. These experiments, however, are based upon thick specimens with 12–25 plies. Due to the constraint of the neighboring layers, an individual ply in the laminate is not free to bend. As a result, modifications to the present theory will be introduced that are necessary before a meaningful comparison with experiments can be made.

It is assumed that the bending-free thermal expansion of a lamina can be realized if there exists a bending moment M_i under a temperature change, ΔT, and no in-plane external force is allowed. Thus,

$$N_i = 0, \qquad \varkappa_i = 0 \tag{1.5-60}$$

From Equation (1.5-8) and $\varkappa_i = 0$:

$$[d^*]\{M\} + \Delta T\{\bar{b}^*\} = 0 \tag{1.5-61}$$

where the subscripts of the matrix quantities are neglected. Then,

$$\{M\} = -\Delta T(d^*)^{-1}\{\bar{b}^*\} \tag{1.5-62}$$

Substituting Equation (1.5-62) into Equation (1.5-8), and from the expression of ϵ^0, a modified in-plane thermal expansion coefficient under the bending-free condition can be defined as:

$$\{a^{**}\} = \}\tilde{a}^*\} - [b^*][d^*]^{-1}\{\bar{b}^*\} \tag{1.5-63}$$

Equation (1.5-63) can be evaluated for the mosaic, crimp, and bridging models provided that the appropriate constants are given for a particular model. Also, note the presence of elastic compliance constants in Equation (1.5-63). Thus, it is necessary to evaluate, for instance, $\bar{b}_{ij}^{*S}$ and $\bar{d}_{ij}^{*S}$ for calculating $\tilde{a}_i^{**S}$, and $\bar{b}_{ij}^{*C}$ and $\bar{d}_{ij}^{*C}$ for $\tilde{a}_i^{**C}$. The above modifications are of practical significance because it is desirable to overcome the anti-symmetrical behavior such as $\bar{b}_i^*$ by suitable stacking in laminate constructions.

Figure 1.5-16 gives the variation of $\tilde{a}_1^{**}$ with $1/n_g$. The theoretical predictions are based upon both crimp and bridging models using the thermoelastic properties of unidirectional graphite/epoxy composite of Table 1.5-2. The experimental results for 5-harness satin composites, which have the same properties as given in Table 1.5-2, are also shown in Figure 1.5-16. Since the thickness of a lamina is not given in reference [20], two estimated values for a/h were used for the analysis and a/a_u is assumed to be unity. The bridging model prediction coincides fairly well with experiments. It is also obvious that the in-plane thermal expansion coefficients are more sensitive to n_g in the bending-free case (Figure 1.5-16) than the bending-unconstrained case (Figure 1.5-14).

1.5.10 Summaries of the Thermal Property Modellings of Two-Dimensional Woven Fabric Composites

1. The mosaic model provides a simple means for estimating thermal expansion and thermal bending coefficients.
2. The one-dimensional crimp model predicts slightly higher in-plane thermal expansion coefficients

and the same thermal bending coefficients as compared to those obtained from the mosaic model. The limited experimental data on thermal bending coefficient coincides favorably with the predictions of the mosaic and crimp models.

3. The bridging model is particularly suited for the prediction of thermal expansion constants for satin composites. The experimental results on in-plane thermal expansion coefficients for a 5-harness satin composite agree well with the theory.

1.5.11 References

1. ISHIKAWA, T. "Anti-Symmetric Elastic Properties of Composite Plates of Satin Weave Cloth," *Fibre Science and Technology*, 15:127 (1981).
2. ISHIKAWA, T. and T. W. Chou. "Elastic Behavior of Woven Hybrid Composites," *Journal of Composite Materials*, 16:2 (1982).
3. ISHIKAWA, T. and T. W. Chou. "Stiffness and Strength Behavior of Woven Fabric Composites," *Journal of Materials Science*, 17:3211 (1982).
4. ISHIKAWA, T. and T. W. Chou. "In-Plane Thermal Expansion and Thermal Bending Coefficients of Fabric Composites," *Journal of Composite Materials*, 17:92–104 (1983).
5. ISHIKAWA, T. and T. W. Chou. "One-Dimensional Micromechanical Analysis of Woven Fabric Composites," *AIAA Journal*, 21:1714 (1983).
6. ISHIKAWA, T. and T. W. Chou. "Nonlinear Behavior of Woven Fabric Composites," *Journal of Composite Materials*, 17:399 (1983).
7. ISHIKAWA, T. and T. W. Chou. "Thermoelastic Analysis of Hybrid Fabric Composites," *Journal of Materials Science*, 18:2260 (1983).
8. ISHIKAWA, T. and T. W. Chou. "Stiffness and Strength Properties of Woven Fabric Composites," *Progress in Science and Engineering of Composite Materials, Proceedings of the Fourth International Conference on Composite Materials* (1982).
9. CHOU, T. W. "Strength and Failure Behavior of Textile Structural Composites," *Proceedings of the American Society for Composites*, 1st Tech. Conf., Technomic Publishing Co., Inc., Lancaster, PA, p. 104 (1986).
10. CHOU, T. W. and J. M. Yang. "Structure-Performance Maps of Polymeric, Metal, and Ceramic Matric Composites," *Metallurgical Transactions*, 17A:1547 (1986).
11. VINSON, J. R. and T. W. Chou. *Composite Materials and Their Use in Structures*. Applied Science Pub., London (1975).
12. JONES, R. M. *Mechanics of Composite Materials*. McGraw-Hill Book Co., New York (1975).

Table 1.5-2. Material properties of a graphite/epoxy unidirectional lamina [20].

E_L	148 GPa
E_T	7.35 GPa
G_{LT}	3.92 GPa
ν_{TL}	0.31
ν_{TT}	0.52
α_L	$3.1 \times 10^{-7} °C^{-1}$
α_T	$3.1 \times 10^{-5} °C^{-1}$

Longitudinal modulus is 237 GPa and transverse modulus is 12 GPa for Grafil XAS fibers.

13. ISHIKAWA, T., K. Koyama and S. Kobayashi. *Journal of Composite Materials*, 11:332 (1977).

14. KIMPARA, I., A. Hamamoto and M. Takehana. *Trans. JSCM*, 3:21 (1977).

15. Kawasaki Heavy Industry, Ltd., private communication.

16. Rhône-Poulenc Corp., "Catalogue of Polyimide Resin."

17. LEKHNITSKII, S. G. *Theory of Elasticity of an Anisotropic Elastic Body*. Holden-Day, San Francisco, CA (1963).

18. ZWEBEN, C. and J. C. Norman. *SAMPE Quarterly*, 1 (July 1976).

19. KIMPARA, J. and M. Takehana. *Proceedings of the 2nd Acoustic Emission Symposium*, Tokyo, Session 9/2-20 (1974).

20. ROGERS, K. F., D. M. Kingston-Lee, L. N. Phillips, B. Yates, M. Chandra and S. F. H. Parker. "The Thermal Expansion of Carbon-Fibre Reinforced Plastics, Part 6, The Influence of Fibre Weave in Fabric Reinforcement," *Journal of Materials Science*, 16:2803 (1981).

21. ROGERS, K. F., L. N. Phillips, D. M. Kingston-Lee, B. Yates, M. J. Overy, J. P. Sargent and B. A. McCalla. "The Thermal Expansion of Carbon Fibre-Reinforced Plastics, Part 1, The Influence of Fibre Type and Orientation," *Journal of Materials Science*, 12:718 (1977).

22. YATES, B., M. J. Overy, J. P. Sargent, B. A. McCalla, D. M. Kingston-Lee, L. N. Phillips and K. F. Rogers. "The Thermal Expansion of Carbon Fibre-Reinforced Plastics, Part 2, The Influence of Fibre Volume Fraction," *Journal of Materials Science*, 13:433 (1978).

23. ZEWI, I. G., I. M. Daniel and J. T. Gotro. "Residual Stresses and Warpage in Woven-Glass/Epoxy Laminates," *Experimental Mechanics*, 27(1):44 (1987).